COVID-19 Impact on Economy, Environment and Healthcare

Alok Raghav • Goo-Bo Jeong
Hamid Ashraf
Editors

COVID-19 Impact on Economy, Environment and Healthcare

Springer

Editors
Alok Raghav
Department of Anatomy and Cell Biology
Gachon University Gil Medical Center
Incheon, Korea (Republic of)

Goo-Bo Jeong
Department of Anatomy and Cell Biology
Gachon University Gil Medical Center
Incheon, Korea (Republic of)

Hamid Ashraf
Aligarh Muslim University
Aligarh, Uttar Pradesh, India

ISBN 978-981-95-1195-2   ISBN 978-981-95-1196-9 (eBook)
https://doi.org/10.1007/978-981-95-1196-9

This Springer imprint is published by the registered company Springer Nature Singapore Pte Ltd.
The registered company address is: 152 Beach Road, #21-01/04 Gateway East, Singapore 189721, Singapore

Preface

The COVID-19 pandemic, a singular and unprecedented global health emergency, has triggered profound and enduring disruptions that transcend the boundaries of virology and epidemiology, permeating the economic, social, political, and environmental fabric of our interconnected world. Evolving from a localized outbreak to a full-scale global crisis, the pandemic exposed deep structural vulnerabilities across nations and sectors. This volume presents a comprehensive, interdisciplinary exploration of COVID-19's far-reaching impact, weaving together insights from epidemiology, economics, healthcare, and environmental science. Initially this book traces the origins and global epidemiological patterns of the outbreak, laying the foundation for understanding its progression and public health implications that further delves deeper into its epidemiological characteristics—highlighting disparities in morbidity, mortality, and healthcare capacity. The ensuing chapters shift focus to the escalating burden of non-communicable and communicable diseases in the post-pandemic era, emphasizing how disruptions in healthcare services and shifts in societal behaviour have intensified existing health challenges. The economic fallout is rigorously analysed in further chapters addressing the recession's disproportionate toll on vulnerable populations, its reverberations across labour markets and trade systems, and the recalibration of fiscal policies in response to economic instability. Environmental consequences also are examined in this book, revealing the transient nature of ecological reprieves during lockdowns and advocating for sustainable recovery strategies. The volume culminates in last chapter with a strategic outlook on future pandemic preparedness, advocating for resilient health infrastructures, equitable economic policies, and integrative environmental stewardship. Designed for public health experts, economists, environmental scholars, policymakers, and students, this book aspires to contribute meaningfully to the global discourse on resilience and recovery, offering both critical reflection and a forward-looking vision for navigating future global health emergencies. Crafted with clarity and accessibility in mind, this volume serves as a valuable resource. The editors extend their profound gratitude to all contributing authors for their scholarly dedication, and to Springer Nature Publishing for their unwavering support and collaboration in bringing this work to successful fruition.

Incheon, Korea (Republic of)	Alok Raghav
Incheon, Korea (Republic of)	Goo-Bo Jeong
Aligarh, Uttar Pradesh, India	Hamid Ashraf

Contents

1 Introduction to COVID-19, Outbreaks, and Its Epidemiology Pattern Worldwide

Savitesh Kushwaha, Rachana Srivastava, and Poonam Khanna

1.1 Introduction

COVID-19, caused by SARS-CoV-2, has resulted in over 6 million deaths globally, marking it as the most significant health crisis since the 1918 influenza pandemic [1]. Originating in Wuhan, China, in December 2019, the virus spread rapidly, leading the WHO to declare a pandemic by March 2020. The causative pathogen was identified as a novel coronavirus, initially named 2019-nCoV, and later renamed SARS-CoV-2 due to its phylogenetic similarity to the coronavirus responsible for the SARS outbreak in 2003. Coronaviruses are a large family of viruses with a crown-like appearance known to infect animals and humans. SARS-CoV and MERS-CoV were previous zoonotic examples that caused severe respiratory illnesses [2]. Coronaviruses are classified into four genera: Alphacoronavirus, Betacoronavirus, Deltacoronavirus, and Gammacoronavirus. SARS-CoV-2, including its variants such as Alpha (B.1.1.7), Beta (B.1.351), Delta (B.1.617.2), Gamma (P.1), and Omicron (B.1.1.529), belongs to the genus Betacoronavirus. SARS-CoV-2 is a novel beta-CoV, sharing a subgenus with SARS-CoV and MERS-CoV, which caused previous epidemics with mortality rates of up to 10% and 35%, respectively [1]. SARS-CoV-2 shares 79% genome sequence identity with SARS-CoV and 50% with MERS-CoV, and it is phylogenetically related to bat coronaviruses [2], sharing 80% nucleotide identity with human SARS-CoV and exhibiting the highest similarity to bat SARS [3].

SARS-CoV-2 adapts through genetic evolution and the development of mutations, just as other RNA viruses. As a result, mutant variations that could differ from their parent strains in some ways are produced. Throughout the epidemic, several SARS-CoV-2 variations have been identified; however, only a small number of these are regarded as variants of concern (VOCs). SARAS-CoV-2 is primarily made up of four structural proteins: spike (S), envelope (E) glycoprotein, nucleocapsid (N), and membrane (M) protein. In addition, it has 5–8 auxiliary proteins and 16 non-structural proteins. The exterior surface of the virion is home to the crown-like glycoprotein known as the surface spike (S) glycoprotein. It cleaves into the amino (N)-terminal S1 subunit, which makes it easier for the virus to enter the host cell. The fusion peptide, transmembrane domain, and cytoplasmic domain found in the carboxyl (C)-terminal S2 subunit cause the virus to fuse its membrane with the cell. The N-terminal domain (NTD), which promotes viral entrance into the host cell and acts as a possible target for neutralization in response to antisera or vaccinations, and the receptor-binding

S. Kushwaha · R. Srivastava · P. Khanna (✉)
Department of Community Medicine and School of Public Health, Postgraduate Institute of Medical Education and Research, Chandigarh, India

A. Raghav et al. (eds.), *COVID-19 Impact on Economy, Environment and Healthcare*,
https://doi.org/10.1007/978-981-95-1196-9_1

domain (RBD) comprise the remaining portion of the S1 subunit. The SARS-CoV-2 spike, or S protein (S1), binds to the ACE2 receptors in the respiratory epithelium, allowing SARS-CoV-2 to enter the host cells (Fig. 1.1). Other tissues that express ACE2 receptors include the upper esophagus, enterocytes from the ileum, cardiac cells, kidney proximal tubular cells, and bladder urothelial cells. After the virus attaches itself, the host transmembrane serine protease 2 (TMPRSS2) primes the spike protein S2 component, facilitating cell entrance and subsequent viral multiplication. Viral replication causes direct tissue damage during the early stages of the infection. The late stage involves the recruitment of T lymphocytes, monocytes, and neutrophils by the infected host cells to start an immune response. Releases include cytokines such as interleukin-1 (IL-1), interleukin-6 (IL-6), granulocyte-macrophage colony-stimulating factor (GM-CSF), tumor necrosis factor-α (TNF-α), IL-1β, IL-8, IL-12, and interferon (IFN)-γ. There is a "cytokine storm" during a severe COVID-19 episode. This is brought on by the immune system being overactive and the bloodstream having a lot of cytokines. Its outcome is both a local and systemic inflammatory reaction [4].

SARS-CoV-2 primarily targets the respiratory system but can also impact other major organ systems, including the gastrointestinal tract (GI), hepatobiliary, cardiovascular, renal, and central neurological systems. Multiple pathways, including direct viral toxicity, thrombosis, vasculitis-induced ischemia injury, immunological dysregulation, and dysregulation of the renin-angiotensin-aldosterone system (RAAS), are likely responsible for SARS-CoV-2-induced organ failure. People of all ages are susceptible to this virus. Patients who are 60 years of age or older, as well as those with underlying medical conditions such as obesity, cardiovascular disease, diabetes, chronic lung disease, smoking, cancer, solid organ transplant recipients, or hematopoietic stem cell recipients, are more likely to experience a severe COVID-19 infection. Data regarding the gender-based differences in COVID-19 shows that male patients have a higher risk of severe illness and increased mortality due to COVID-19 compared to female patients [5, 6].

1.2 Symptoms of COVID-19 and Beyond

The severe acute respiratory syndrome coronavirus 2 (SARS-CoV-2) that produced the COVID-19 pandemic has posed a serious threat to healthcare systems around the globe. Although the acute stage of the illness has received much attention, a

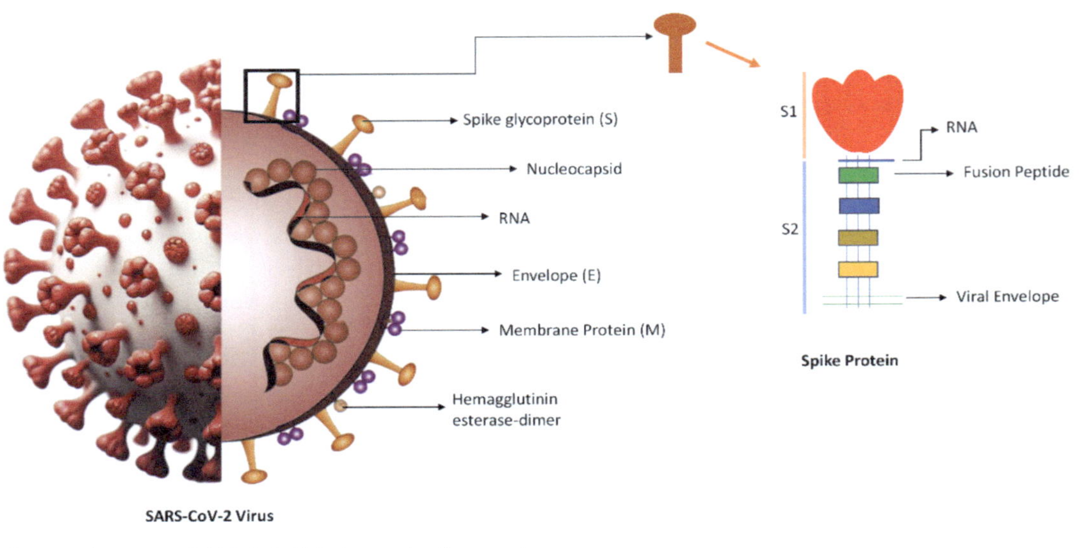

Fig. 1.1 Structure of SARS-CoV-2 virus and spike protein

growing corpus of research focuses on the ongoing symptoms that COVID-19 patients endure beyond the acute phase. 97.5% of symptomatic people developed symptoms after 11.5 days of exposure, with an average incubation time of 5 days for the virus. The most frequently reported symptoms are dyspnea, fever, and dry cough. Other, less common symptoms have also been noted, and the presentation of symptoms varies greatly [7]. Although these are not exclusive to COVID-19, radiographic and laboratory results frequently show lymphopenia and increased lactate dehydrogenase levels in individuals. With a mean incubation period of 5.2 days (95% confidence interval, 4.1–7.0), COVID-19 symptoms range from mild to severe. The most typical symptoms are fever (83–98%), cough (76–82%), and shortness of breath (31–55%) [8]. Reverse transcription polymerase chain reaction (RT-PCR) testing is the main method used to make the diagnosis; nevertheless, false-negative results can occur in 20–67% of patients, depending on the test's quality and timing. Approximately 5% of individuals and 20% of hospitalized patients have severe cases of COVID-19, which can result in abrupt respiratory failure and the requirement for urgent care. More than 75% of hospitalized COVID-19 patients require supplementary oxygen, making its use ubiquitous [7]. Emerging therapies like remdesivir have been discovered to shorten the time to recovery, and dexamethasone has demonstrated a decrease in the 28-day mortality rates for patients needing extra oxygen [7]. Furthermore, a prothrombotic condition and a significant frequency of thrombotic problems in critically ill patients have been linked to COVID-19. According to autopsy reports, 71% of COVID-19-related deaths had diffuse intravascular coagulation (DIC). Even in the absence of a vaccination that works, preventive measures are still essential. These consist of contact tracing, social distance, and the usage of face masks [9].

Elliott et al. (2021) reported that 6450 positive cases were found using PCR testing out of 1,147,370 individuals in the REACT-1 trial who were 5 years of age or longer. Between 22% and 27% of participants completed the trial between June 2020 and January 2021 [10]. PCR-positive samples with an N-gene cycle threshold value <34 were subjected to viral genome sequencing. Participants reported symptoms they had experienced in the week before testing. When compared to people who had no symptoms, all 26 survey symptoms were linked to PCR positivity in univariate analysis. However, seven symptoms were found to be jointly predictive of PCR positivity using stability selection, a technique that involved 1000 penalized logistic regression models with 50% subsampling: fever, chills, new persistent cough, loss of appetite, muscle aches, and loss or change of smell or taste. The results of the REACT-1 study indicate that utilizing these seven symptoms to determine test allocation could lead to a higher number of tests conducted, with as many as 30–40% of symptomatic persons undergoing testing, allowing for the detection of up to 75% of symptomatic cases [10]. This strategy would optimize case discovery within the community, particularly when testing capacity is constrained. Most patients who recover from COVID-19 still have symptoms, with fatigue and dyspnea being particularly common [11]. The necessity of recognizing clinical signs for an early diagnosis is underscored by the lack of specific therapy for the virus and the scarcity of critical care units and mechanical ventilation globally [12]. A systematic study is especially pertinent in light of the worldwide diagnostic kit shortage [13]. The meta-analysis's conclusions align with the typical signs and symptoms of viral infections. The most typical symptom is fever, albeit not every patient has it [14]. The fever is linked to more severe illness and a longer hospital stay, combined with vomiting and a high temperature (over 39 degrees) [15]. According to the Covid Symptoms Study, sore throat, headache, exhaustion, runny nose, and sneezing were the top five symptoms of Omicron infection (Table 1.1). This information was obtained from positive cases in London at a time when the prevalence of Omicron was high [16]. When 363 Delta instances from 3 to 10 October 2021 and 847 Omicron cases from 3 to 10 December 2021 were analyzed, the study did not find any statistically significant variations in the early symptoms reported three days after testing. The fact that the initial symptoms of Delta and Omicron are similar is significant

Table 1.1 List of COVID-19 symptoms

Common symptoms of COVID-19	
Loss or change of sense of smell	Chest pain
Loss or change of sense of taste	Gastrointestinal symptoms
Fever	Arthritis
Persistent cough	Anorexia
Chills	Allergy-like symptoms
Appetite loss	Dyssomnias
Muscle aches	Ear-related problems
Fatigue	Eye-related problems
Dyspnea (difficulty breathing)	Skin problems
Sputum production	Dehydration
Headache	Anemia
Weakness	Urination problems
Nausea/vomiting	Enlargement of lymph nodes
Diarrhea	Sinus enlargement
Chest tightness	Oral problems

because it indicates that COVID-19's symptomatology—regardless of variant—may be moving toward a more unified appearance. Nonetheless, there have been requests for the UK government to revise its list of COVID-19 symptoms to better represent the current clinical picture, given the frequency of cold-like symptoms in Omicron cases. In contrast to other health authorities, the UK is reluctant to update the symptom list. The World Health Organization, the US Centers for Disease Control and Prevention, and European nations, including France and Spain, have all revised their recommendations to add cold-like symptoms as probable markers of COVID-19 infection [16].

The post-COVID-19 condition (PCC), sometimes referred to as "long COVID," is one of the disease's most worrisome features. It is characterized by persistent symptoms that last long after the infection's acute phase. Long COVID is characterized by persistent symptoms that occur 90–365 days after a confirmed COVID-19 infection, without a history of these symptoms 180 days before infection. The relationships between anxiety, depressive symptoms, and post-COVID-19 condition (PCC) were explored in people in France through a national survey. 1095 individuals with PCC and 1021 SARS-CoV-2-infected individuals without PCC were included in the study. It was discovered that 21% and 18% of people self-reported having anxiety or depression, respectively, and that 33% and 20% of people now exhibited measured symptoms of these conditions. Nevertheless, only 7.6% of depressed symptoms and 13.4% of anxiety symptoms satisfied the World Health Organization's (WHO) standards for PCC [17]. Another study found a positive relationship between the number of physical symptoms experienced in the hospital and the number of symptoms reported at T1 and T2 [14]. The disease's early intensity may indicate how long symptoms will last after discharge. Previous studies have shed light on the possible long-term consequences of COVID-19 by examining the psychological and physiological implications of critical illness, such as Post Intensive Care Syndrome (PICS). Financial hardships and problems accessing mental health services are reported by PICS patients, which may also apply to PASC (Post-Acute Sequelae of SARS-CoV-2) patients [18].

1.3 Transmission Dynamics

When people are in close proximity to one another, respiratory droplets are the main way that SARS-CoV-2 can spread (Fig. 1.2). The infection can spread through asymptomatic, presymptomatic, and symptomatic carriers. Although there is a chance of transmission through inanimate surfaces, droplets are thought to be the main method of transmission [9]. Increased residential time resulting from mobility constraints positively affected subsequent infections [19]. A 5% increase in residential time was associated with an 82.5% reduction in cumulative infections ten days later. It has been determined that vaccination is a more effective way to stop the spread of viruses [20]. Nearly 13 billion vaccination doses had been given worldwide as of November 29, 2022, and the World Health Organization (WHO) is promoting immunization rates of at least 70% in every nation [21]. According to estimates, each infected person in a susceptible population can infect two to three

Fig. 1.2 Types of transmissions of SARS-CoV-2

others, with the basic reproduction number (R0) of COVID-19 ranging from 2 to 3 [22, 23]. However, the pandemic's path has been greatly impacted by super-spreading occurrences, in which a small number of people infect a disproportionately large number of contacts [24]. The spread of COVID-19 has been significantly aided by asymptomatic transmission. According to studies, those who don't exhibit any symptoms may nevertheless have a viral load similar to that of symptomatic patients, which could lead to silent transmission chains [25]. Using face masks could significantly lower the probability of COVID-19 transmission, with an odds ratio of 0.15 in favor of mask use [26]. Social distancing measures have also been linked to a decline in the effective reproduction number (Rt), with a median decline of 13% for every extra meter of separation [27]. Data from the real world have also demonstrated a significant effect on transmission, with vaccinated people showing lower virus loads and a decreased chance of spreading the infection to others [28]. Mutations that boost transmissibility and may partially evade immune responses are seen in variants of concern, such as B.1.1.7 (Alpha) and B.1.617.2 (Delta) [29].

Specifically, the Delta variation has demonstrated a greater R0 and has been linked to a higher risk of hospitalization [30].

The environment has an impact on transmission as well. According to a meta-analysis, there may be a seasonal trend because every 1 °C increase in temperature was linked to a 3.1% decrease in daily COVID-19 instances [31]. Lower humidity and temperatures, on the other hand, may increase the likelihood of virus propagation by extending its viability [32]. Temperature and the frequency of COVID-19 cases and deaths were found to be negatively correlated in a study involving 16 member states of the African Union [32]. Higher air temperatures significantly reduce COVID-19 transmission, according to an analysis of meteorological data from the National Oceanic and Atmospheric Administration (NOAA). These findings are consistent with those suggesting a seasonal pattern in European countries [33, 34]. But human actions and intervention efforts probably outweigh the effects of climate change [35].

Studies have indicated that air travel greatly increases the spread of SARS [36]. For COVID-19, comparable trends have been noted, with flight volume highly correlated with the number of confirmed cases [23, 37]. The transmission of COVID-19 was also significantly aided by railway travel; cities connected to Wuhan via high-speed rail had 40% higher infection rates [23, 38, 39]. Although less extensively studied, road travel has been associated with the transmission of COVID-19 and H1N1 [40–42]. According to published research, there is a correlation between increased case and mortality rates and proximity to transit hubs and airports [43, 44]. Moreover, the persisting effects of COVID-19 on transportation facilities indicate that safety and mobility-enhancing measures will be more durable [45].

Governments have taken several measures to stop the virus's spread, including lockdowns and social distancing. Respiratory droplets and contact with contaminated surfaces are the main ways that SARS-CoV-2 is spread. For 1 week, the virus can infect materials such as metal, wood, and glass [46, 47]. The epidemic has forced global organizational changes, including expanded hospital capacity, remote work and learning, and postponing non-COVID-19 treatments.

1.4 Diagnosis and Detection Techniques

Several diagnostic tests have had to be developed and implemented in response to the COVID-19 pandemic to detect and contain the spread of SARS-CoV-2. The kind of specimen used is essential to the COVID-19 test's accuracy. Nasopharyngeal swabs (NPS), one type of upper respiratory specimen that is frequently utilized and has been demonstrated to produce accurate results, are one example. Sputum and other lower respiratory specimens are advised during the later stages of the illness or if upper respiratory specimens yield negative results due to their greater virus loads [48]. Although blood specimens are not as frequently used for diagnosis, the presence of the virus in plasma can suggest a serious illness. When respiratory specimens are negative, fecal specimens may be considered as they have been linked to severe disease [49].

The gold standard for identifying an active COVID-19 infection is an NAAT (nucleic acid amplification tests), especially real-time polymerase chain reaction (RT-PCR) assays [50]. These tests need specialized workers and laboratory equipment, but they also have a high sensitivity and specificity. Sample-to-answer molecular diagnostics platforms are more expensive and less appropriate for mass screening, but they provide quick findings with less operator involvement. With an average sensitivity of 56% and a high specificity of over 97%, antigen detection tests are less sensitive than NAATs [49]. These assessments are helpful when assessing patients early in their illness and in environments where access to NAATs is restricted. When NAATs are unavailable, the World Health Organization (WHO) recommends using antibody detection tests within the first 5–7 days after the onset of symptoms. These tests target the SARS-CoV-2 nucleocapsid (N) or spike (S) proteins and vary in the antibodies measured, with IgG or total immunoglobulin assays having higher accuracy than IgM or IgA assays. One concern with antibody detection tests is cross-reactivity with other coronaviruses, as they can be used to diagnose acute infection when a change in antibody titers is detected [49]. The POCTs (point-of-care tests) can be used outside laboratories and are easy to use, but their sensitivity may be lower than that of laboratory tests. For POCT platforms, technologies such as rolling circle amplification (RCA), nicking enzyme-assisted reaction (NEAR), recombinase polymerase amplification (RPA), and loop-mediated isothermal amplification (LAMP) have been adopted [51]. Developing human antibodies 2–3 weeks after infection can produce false-positive results even with effective immunological detection methods. Nanomaterial-based biosensors have been developed to increase sensitivity and specificity. The detecting platforms these sensors generate are affordable thanks to their utilization of nanotechnology advancements. As a substitute, RAT (Rapid Antigen Test) kits have become popular since they are simple to administer at the point of care and provide quick results. The study revealed that the RAT kits had an overall pooled specificity of 99.4% (95% CI: 99.1–99.8; $I^2 = 0.90$) and sensitivity of 68.4% (95% CI: 60.8–75.9; $I^2 = 98\%$). These results imply that in contrast to RT-PCR, RAT kits have a reduced sensitivity even when highly specific. Test performance can be influenced by the kind of specimen and patient symptoms, as seen by the increased sensitivity of RAT kits in nasopharyngeal specimens and symptomatic patients [52]. As the World Health Organization specified, the minimal performance requirements for RAT kits are 80% sensitivity and 97% specificity. Even while some RAT kits satisfy these requirements, the sensitivity variations between studies emphasize the necessity of ongoing research and development to enhance RAT kit functionality. The alarming conclusion of a study included the possibility that antigen-based testing may provide negative test results for 60% of patients infected with SARS-CoV-2, which could result in subpar clinical care and a rise in viral transmission [53]. This sensitivity surpasses the sensitivity of 30.2% discovered in a Belgian study [54]. The Mexican study's decision to include symptomatic patients exclusively may have contributed to the disparity by raising the pre-test risk of infection and, thus, its sensitivity [53]. Further development and optimization of these technologies are needed to address the upcoming pandemic.

1.5 COVID-19 Events: Cases, Mortality, and Vaccines

Events	Date
Pneumonia-like illness in Wuhan, China.	12-12-2019
First report of 27 cases of pneumonia with unknown cause in Wuhan, China.	31-12-2019
WHO got informed of several cases.	31-12-2019
Public health officials in China identify a novel coronavirus as the causative agent of the outbreak.	07-01-2020
WHO begins using the phrase "2019 Novel Coronavirus" or "2019-nCoV" to refer to disease causing the outbreak in Wuhan, China.	10-01-2020
Four other genomes have been uploaded to the viral sequence database curated by the Global Initiative on Sharing All Influenza Data (GISAID).	11-01-2020
Thailand confirms the first laboratory-confirmed case of the SARS-CoV-2 virus outside of China.	13-01-2020
WHO finds evidence of possible human-to-human transmission of the SARS-CoV-2 virus.	14-01-2020
FDA approves the EUA for the CDC-developed SARS-CoV-2 diagnostic test kit.	04-02-2020
Worldwide deaths from the 2019 Novel Coronavirus reach 1013.	10-02-2020
WHO announces the name for the disease causing the 2019 Novel Coronavirus outbreak: "COVID-19."	11-02-2020
Italy has become a global COVID-19 hotspot.	23-02-2020
After more than 118,000 cases in 114 countries and 4291 deaths, the WHO declares COVID-19 a pandemic.	11-03-2020
US bans non-US citizens from EU countries.	13-03-2020
Moderna Therapeutics begin the first human vaccine trials to protect against COVID-19 at a research facility in Seattle, Washington.	17-03-2020
WHO published a draft landscape of COVID-19 candidate vaccines.	11-04-2020
FDA issues an emergency use authorization (EUA) for using the antiviral drug Remdesivir to treat suspected or confirmed COVID-19 in people hospitalized with severe disease.	01-05-2020
AstraZeneca receives more than $1 billion from the U.S. government in funding for developing the AstraZeneca/Oxford University COVID-19 vaccine.	21-05-2020
WHO announced that the hydroxychloroquine arm of the Solidarity Trial to find an effective COVID-19 treatment was being stopped. Hydroxychloroquine did not reduce mortality for hospitalized COVID-19 patients.	17-06-2020
WHO announces that the SARS-CoV-2 virus that causes COVID-19 can be transmitted through the air and is likely spread by asymptomatic individuals.	09-07-2020
Moderna, the first potential coronavirus vaccine tested in humans, shows a positive immune response.	14-07-2020
Obesity is found to increase the risk of mortality from COVID-19 disease, especially among men and younger people—even when isolated from racial, ethnic, or socioeconomic disparities.	12-08-2020
Russia has begun production of the Sputnik V COVID-19 vaccine.	15-08-2020
The University of Hong Kong confirms the first documented case of COVID-19 reinfection.	24-08-2020
India has become the country with the third-highest COVID-19 cases, behind the United States and Brazil.	31-08-2020
The Journal of the American Medical Association and WHO now recommend the use of steroids for the treatment of severe COVID-19.	03-09-2020
According to Johns Hopkins University, global deaths top 900,000	10-09-2020
The global coronavirus death toll has topped 1 million.	27-09-2020
The world has passed 50 million coronavirus cases.	08-11-2020
The United Kingdom grants the Pfizer-BioNTech vaccine candidate the world's first emergency use authorization.	02-12-2020
The United Kingdom identified a new variant of SARS-CoV-2 that appeared to be more transmissible (Alpha B.1.1.7).	08-12-2020
The United States grants the Pfizer-BioNTech candidate its first emergency use authorization for a COVID-19 vaccine.	11-12-2020
After the United States, India became the second country to record 10 million COVID-19 cases.	18-12-2020
Global COVID-19 cases surpass 80 million.	26-12-2020

(continued)

Events	Date
The Pfizer-BioNTech vaccine has been reported as effective against a mutation in new COVID-19 variants.	07-01-2021
UNICEF announces a deal with India's Serum Institute to produce 1.1 billion doses of COVID-19 vaccine for low-income countries.	03-02-2021
The AstraZeneca vaccine is effective against the coronavirus variant first found in the United Kingdom.	04-02-2021
The World Health Organization reports its sixth consecutive week of declining COVID-19 cases globally.	23-02-2021
Researchers find a new COVID-19 variant in New York City (B.1.5.2.6).	25-02-2021
AstraZeneca finds no risk of increased blood clotting with its COVID-19 vaccine.	14-03-2021
The European Commission proposes an EU vaccine passport: the Digital Green Certificate. The United Kingdom considered introducing vaccine passports for summer travel.	17-03-2021
WHO released its report on the origins of SARS-CoV-2, finding that animal-to-human transmission was the most likely origin of the virus.	30-03-2021
Pfizer and BioNTech announced their coronavirus vaccine is extremely effective in adolescents 12–15 years old.	31-03-2021
European Medicines Agency finds possible link between AstraZeneca's COVID-19 vaccine and very rare cases of blood clots.	04-04-2021
The United States paused using the Johnson & Johnson vaccine after six individuals (out of nearly seven million recipients) developed a rare clotting disorder.	13-04-2021
Global coronavirus deaths surpass 3 million.	19-04-2021
India records more than 310,000 new infections in a single day, surpassing the global record for one-day high of recorded cases set by the United States in January.	22-04-2021
Brazilian authorities rejected Russia's Sputnik V vaccine, stating that the doses received from Russia still had replication-competent adenovirus.	27-04-2021
The United States announced $100 million in aid for India.	28-04-2021
Global coronavirus cases surpass 150 million.	30-04-2021
WHO approves Moderna for emergency use.	30-04-2021
India becomes the first country to record more than 400,000 cases daily.	01-04-2021
G20 Tourism ministers support vaccine passports to resume tourism and enable economic recovery.	04-05-2021
The European Medicines Agency begins a rolling review of China's Sinovac vaccine.	04-05-2021
Russia authorizes using the vaccine Sputnik Light, a one-shot dose still undergoing testing.	06-05-2021
The United Kingdom recommends that individuals under 40 be offered an alternative to AstraZeneca when possible.	07-05-2021
An Indian government panel recommends widening the gap between the first and second doses of AstraZeneca to expand vaccine access.	13-05-2021
CureVac's vaccine candidate falls short of 50% efficacy in interim clinical trial results.	16-06-2021
WHO announces that the Delta variant is likely to become globally dominant.	18-06-2021
The WHO recommends that even fully vaccinated people keep wearing masks and social distancing to combat the Delta variant.	25-06-2021
Moderna says that its vaccine produces protective antibodies against the Delta variant.	29-06-2021
The EU Digital COVID Certificate goes live in Europe.	01-07-2021
The first case of COVID-19 is detected in Tokyo's Olympic village.	17-07-2021
USAID announces $720 million for COVID-19 response and recovery globally.	04-08-2021
The Food and Drug Administration grants full approval to the Pfizer-BioNTech vaccine.	23-08-2021
The U.S. intelligence community finds that spillover and lab accident theories are plausible for COVID-19 origins.	27-08-2021
The United States authorizes the Pfizer vaccine for children from 5 to 11 years old.	29-10-2021
Johns Hopkins University reports that the coronavirus has killed more than five million people in less than 2 years.	01-11-2021
The World Health Organization grants emergency use listing for Bharat Biotech's Covaxin vaccine.	03-11-2021

(continued)

Events	Date
The World Health Organization classifies Omicron as a variant of concern.	26-11-2021
A new variant of Omicron BA.2 emerges in Asia and Europe.	24-01-2022
Global COVID-19 death toll surpasses 6 million.	07-03-2022
Pfizer says three doses of its COVID-19 vaccine produce a strong immune response in young children.	23-05-2022
A New Lancet study suggests delta strain is more likely to cause long COVID than omicron.	17-06-2022
WHO Director-General urges Pfizer to make COVID-19 antiviral treatments available in low- and middle-income countries.	06-07-2022
The world has reached the tragic milestone of one million COVID-19 deaths so far in 2022.	22-08-2022
India and China approved two new needle-free COVID-19 vaccines, one administered nasally and the other orally.	06-09-2022
The U.S. Food and Drug Administration (FDA) approves the COVID-19 bivalent booster produced by Pfizer-BioNTech and Moderna for children between the ages of five and eleven.	12-10-2022
Chinese officials declare a "major decisive victory" against COVID-19.	17-02-2023
U.S. President Joe Biden signs a bill ending the COVID-19 national emergency.	10-03-2023

1.6 The Outbreak

Beginning in Wuhan, China, in December 2019, the pandemic extended to 210 nations by April 10, 2020, with different reports of cases and fatalities (Fig. 1.3). Global healthcare systems were severely strained by the quick expansion and lack of preparation, especially in nations with high incidence rates like China, Iran, and Italy [55]. Given that China's economy is now significantly larger than that of the United States, the impact of COVID-19 is anticipated to be far greater than the estimated $40 billion in worldwide economic damages caused by the SARS pandemic of 2002–2003. The disruption of global supply networks, the increase in unemployment, industrial recessions, and decrease in several sectors, including agriculture and education, were observed globally. As of the publication date, 190,000 individuals had perished, and 270 million people had contracted COVID-19, according to data from Johns Hopkins University [56].

Since no particular antiviral medication has shown to be successful, supportive care is the mainstay of treatment [57]. According to early epidemiological trends, COVID-19 vulnerability may be related to biological sex, age, and other medical factors, such as the coronaviruses that cause SARS and MERS. Unprecedented responses to the epidemic included lockdowns and border closures, which significantly increased unemployment. In South Korea, for example, the unemployment rate went from 3.0% to 3.8% [56]. Sales in the retail industry fell precipitously, with year-over-year fluctuations indicating a notable downturn in nations such as China and the United States. The estimate that China's GDP growth would reduce by 4.5% in the first quarter of 2020, with a worldwide GDP decline of around 0.42%, further demonstrated the pandemic's influence on the world economy [56]. The nations closest to Italy's track exhibited power law growth, whereas those with more recent outbreaks tended to display exponential growth. This points to a global pattern of infection that begins exponentially and eventually takes on characteristics of a power law [58].

Online and mobile GIS apps offered near-real-time tracking of the virus's progress during the COVID-19 epidemic, assisting international efforts to monitor and contain the pandemic. One important information source was the Johns Hopkins University CSSE dashboard, which provided interactive maps to find and count confirmed infections, deaths, and recoveries. The dashboard used information from reliable sources, including the National Health Commission of the People's Republic of China, the US Centers for Disease Control and Prevention, and the World Health Organization [59]. The WHO also unveiled its dashboard, offering a worldwide perspective of the out-

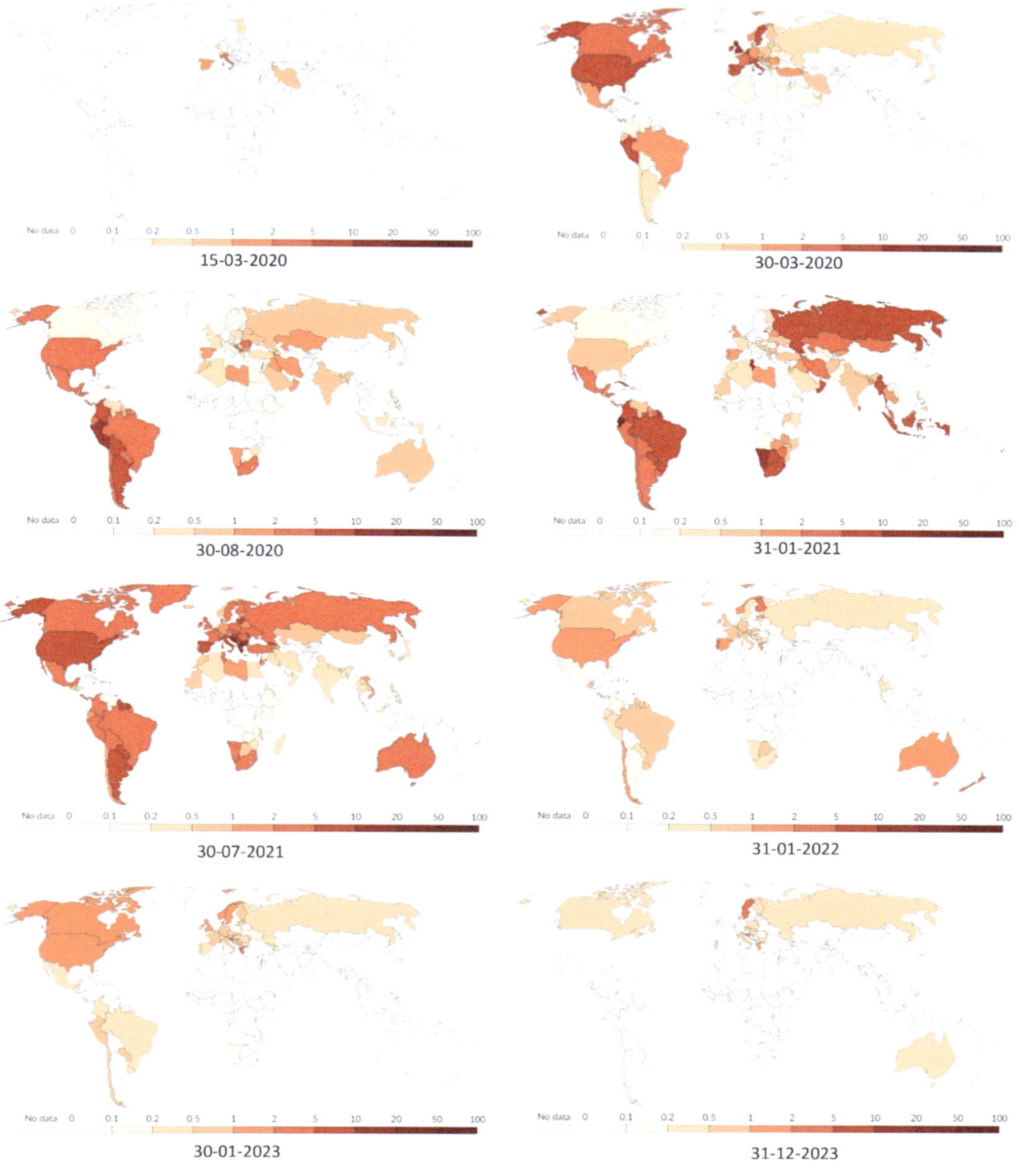

Fig. 1.3 Global timeline map of cumulative COVID-19 spread per million people relative to population. (Source: ourworlddata.org, licensed under CC-BY 4.0)

break and laboratory-confirmed cases broken down by Chinese province and nation.

Throughout 2020 and 2021, the world faced unprecedented challenges due to the COVID-19 pandemic. Key developments during this period included:

- Global Lockdowns and Public Health Measures: To stop the virus's spread, nations worldwide put in place several measures, including mask laws, social separation, and lockdowns.

- Transmission Patterns and Variants: The virus showed a high capacity for mutation, which resulted in the formation of variants that presented further difficulties in controlling the transmission. These variations, which included Delta, Beta, and Alpha, displayed variations in vaccination effectiveness and transmissibility.
- Vaccine Research and Development: The development and approval of multiple COVID-19 vaccines for emergencies resulted from an incredible international effort. Global vaccination campaigns got underway by the end of 2020 and picked up speed in 2021, providing optimism for a possible escape from the pandemic.
- Global Response: The unprecedented cooperation in vaccine development, the exchange of public health tactics, and financial support comprised the global response.

The development of variations was one of the recurring difficulties and adjustments brought about by COVID-19 in 2021–2022, necessitating revisions to vaccination plans and public health recommendations. The introduction of variations such as Delta and Omicron forced the government to modify policies on mask use, social distancing, and lockdown procedures. A significant change in vaccination programs was the recommendation to improve protection against these variations using booster injections. During these times, the burden on international health systems was highlighted as numerous nations struggled with vaccine distribution, healthcare capability, and economic stability. With an emphasis on minimizing the effects on health systems and the economy worldwide, the adjustments to vaccination programs and public health regulations highlighted the dynamic character of the pandemic response.

The COVID-19 pandemic of 2023–2024 has shown trends toward endemicity, with nations adjusting to control the virus and providing standard healthcare. Worldwide vaccination rates have increased, which has helped reduce hospital stays and serious cases. However, new varieties challenge these efforts, calling for constant observation, vaccination updates, and booster shots. To reduce the virus's negative effects on economies and public health, the emphasis has switched to incorporating COVID-19 management into standard healthcare systems. Resilience, adaptability, and the significance of international collaboration in research and health are highlighted now.

From its beginning until its transition into an endemic phase, the COVID-19 pandemic has shown the significance of international collaboration, scientific advancement, and flexibility in public health approaches. The world has made great progress in lessening the effects of the virus, even if issues with managing variations and vaccine delivery still exist. As we progress toward a future when COVID-19 is controlled as part of our global health environment, the future path highlights resilience and the need for ongoing monitoring.

References

1. Cascella M, Rajnik M, Aleem A, et al. Features, evaluation, and treatment of coronavirus (COVID-19). Treasure Island: StatPearls Publishing Copyright © 2024, StatPearls Publishing LLC; 2024.
2. Zhou H, Yang J, Zhou C, et al. A review of SARS-CoV2: compared with SARS-CoV and MERS-CoV. Front Med. 2021;8:628370.
3. Purnell TS, Simpson DC, Callender CO, et al. Dismantling structural racism as a root cause of racial disparities in COVID-19 and transplantation. Am J Transplant. 2021;21:2327–32.
4. Cascella M, Del Gaudio A, Vittori A, et al. COVID-pain: acute and late-onset painful clinical manifestations in COVID-19—molecular mechanisms and research perspectives. J Pain Res. 2021;14:2403–12.
5. Gebhard C, Regitz-Zagrosek V, Neuhauser HK, et al. Impact of sex and gender on COVID-19 outcomes in Europe. Biol Sex Differ. 2020;11:29.
6. Jin JM, Bai P, He W, et al. Gender differences in patients with COVID-19: focus on severity and mortality. Front Public Health. 2020;8:152.
7. Wiersinga WJ, Rhodes A, Cheng AC, et al. Pathophysiology, transmission, diagnosis, and treatment of coronavirus disease 2019 (COVID-19): a review. JAMA. 2020;324:782–93.
8. Wu Y-C, Chen C-S, Chan Y-J. The outbreak of COVID-19: an overview. J Chin Med Assoc. 2020;83:217–20.
9. Wiersinga WJ, Prescott HC. What is COVID-19? JAMA. 2020;324:816.

10. Elliott J, Whitaker M, Bodinier B, et al. Predictive symptoms for COVID-19 in the community: REACT-1 study of over 1 million people. PLoS Med. 2021;18:e1003777.
11. Carfì A, Bernabei R, Landi F. Persistent symptoms in patients after acute COVID-19. JAMA. 2020;324:603–5.
12. Huang C, Wang Y, Li X, et al. Clinical features of patients infected with 2019 novel coronavirus in Wuhan, China. Lancet. 2020;395:497–506.
13. Alimohamadi Y, Sepandi M, Taghdir M, et al. Determine the most common clinical symptoms in COVID-19 patients: a systematic review and meta-analysis. J Prev Med Hyg. 2020;61:E304–e12.
14. Wang C, Pan R, Wan X, et al. A longitudinal study on the mental health of general population during the COVID-19 epidemic in China. Brain Behav Immun. 2020;87:40–8.
15. Guan Q, Sadykov M, Mfarrej S, et al. A genetic barcode of SARS-CoV-2 for monitoring global distribution of different clades during the COVID-19 pandemic. Int J Infect Dis. 2020;100:216–23.
16. Iacobucci G. Covid-19: runny nose, headache, and fatigue are commonest symptoms of omicron, early data show. BMJ. 2021;375:n3103.
17. Tebeka S, Carcaillon-Bentata L, Decio V, et al. Complex association between post-COVID-19 condition and anxiety and depression symptoms. Eur Psychiatry. 2023;67:e1.
18. Effiong A. Postacute sequelae of COVID-19 and adverse psychiatric outcomes: protocol for an etiology and risk systematic review. JMIRx Med. 2023;4:e43880.
19. Ilin C, Annan-Phan S, Tai XH, et al. Public mobility data enables COVID-19 forecasting and management at local and global scales. Sci Rep. 2021;11:13531.
20. Lin C, Lau AKH, Fung JCH, et al. A mechanism-based parameterisation scheme to investigate the association between transmission rate of COVID-19 and meteorological factors on plains in China. Sci Total Environ. 2020;737:140348.
21. Bell LV, Fitzgerald SF, Flusk D, et al. Healthcare provider knowledge, beliefs, and attitudes regarding opioids for chronic non-cancer pain in North America prior to the emergence of COVID-19: a systematic review of qualitative research. Can J Pain. 2023;7:2156331.
22. Liu CY, Berlin J, Kiti MC, et al. Rapid review of social contact patterns during the COVID-19 pandemic. Epidemiology. 2021;32:781–91.
23. Liu J, Xie W, Wang Y, et al. A comparative overview of COVID-19, MERS and SARS: review article. Int J Surg. 2020;81:1–8.
24. Althouse BM, Wenger EA, Miller JC, et al. Superspreading events in the transmission dynamics of SARS-CoV-2: opportunities for interventions and control. PLoS Biol. 2020;18:e3000897.
25. Oran DP, Topol EJ. Prevalence of asymptomatic SARS-CoV-2 infection: a narrative review. Ann Intern Med. 2020;173:362–7.
26. Chu H, Chan JF, Yuen TT, et al. Comparative tropism, replication kinetics, and cell damage profiling of SARS-CoV-2 and SARS-CoV with implications for clinical manifestations, transmissibility, and laboratory studies of COVID-19: an observational study. Lancet Microbe. 2020;1:e14–23.
27. Jones RC, Ho JC, Kearney H, et al. Evaluating trends in COVID-19 research activity in early 2020: the creation and utilization of a novel open-access database. Cureus. 2020;12:e9943.
28. Levine-Tiefenbrun M, Yelin I, Alapi H, et al. Viral loads of Delta-variant SARS-CoV-2 breakthrough infections after vaccination and booster with BNT162b2. Nat Med. 2021;27:2108–10.
29. Davies NG, Klepac P, Liu Y, et al. Age-dependent effects in the transmission and control of COVID-19 epidemics. Nat Med. 2020;26:1205–11.
30. Gunadi, Hakim MS, Wibawa H, et al. Is the infection of the SARS-CoV-2 delta variant associated with the outcomes of COVID-19 patients? Front Med (Lausanne). 2021;8:780611.
31. Byun WS, Heo SW, Jo G, et al. Is coronavirus disease (COVID-19) seasonal? A critical analysis of empirical and epidemiological studies at global and local scales. Environ Res. 2021;196:110972.
32. Mwiinde AM, Siankwilimba E, Sakala M, et al. Climatic and environmental factors influencing COVID-19 transmission—an African perspective. Trop Med Infect Dis. 2022;7(12):433.
33. Smith RB, Mahnert ND, Foote J, et al. Mindfulness effects in obstetric and gynecology patients during the coronavirus disease 2019 (COVID-19) pandemic: a randomized controlled trial. Obstet Gynecol. 2021;137:1032–40.
34. Zoran MA, Savastru RS, Savastru DM, et al. Cumulative effects of air pollution and climate drivers on COVID-19 multiwaves in Bucharest, Romania. Process Saf Environ Prot. 2022;166:368–83.
35. Baker D, Amor S, Kang AS, et al. The underpinning biology relating to multiple sclerosis disease modifying treatments during the COVID-19 pandemic. Mult Scler Relat Disord. 2020;43:102174.
36. Rosca EC, Heneghan C, Spencer EA, Brassey J, Plüddemann A, Onakpoya IJ, Evans DH, Conly JM, Jefferson T. Transmission of SARS-CoV-2 associated with aircraft travel: a systematic review. J Travel Med. 2021;28(7):133. https://doi.org/10.1093/jtm/taab133. PMID: 34480171; PMCID: PMC8499932.
37. Liu A, Kim YR, O'Connell JF. COVID-19 and the aviation industry: the interrelationship between the spread of the COVID-19 pandemic and the frequency of flights on the EU market. Ann Tour Res. 2021;91:103298.
38. Zhao S, Lin Q, Ran J, et al. Preliminary estimation of the basic reproduction number of novel coronavirus (2019-nCoV) in China, from 2019 to 2020: a data-driven analysis in the early phase of the outbreak. Int J Infect Dis. 2020;92:214–7.
39. Zhao S, Musa SS, Lin Q, et al. Estimating the unreported number of novel coronavirus (2019-nCoV)

cases in China in the first half of January 2020: a data-driven modelling analysis of the early outbreak. J Clin Med. 2020;9(2):388.
40. Zhang L, Yang H, Wang K, et al. Measuring imported case risk of COVID-19 from inbound international flights—a case study on China. J Air Transp Manag. 2020;89:101918.
41. Zhang Y, Zhang A, Wang J. Exploring the roles of high-speed train, air and coach services in the spread of COVID-19 in China. Transp Policy (Oxf). 2020;94:34–42.
42. Lu R, Zhao X, Li J, et al. Genomic characterisation and epidemiology of 2019 novel coronavirus: implications for virus origins and receptor binding. Lancet. 2020;395:565–74.
43. Gaskin DJ, Zare H, Delarmente BA. Geographic disparities in COVID-19 infections and deaths: the role of transportation. Transp Policy (Oxf). 2021;102:35–46.
44. Chokshi A, DallaPiazza M, Zhang WW, et al. Proximity to international airports and early transmission of COVID-19 in the United States—an epidemiological assessment of the geographic distribution of 490,000 cases. Travel Med Infect Dis. 2021;40:102004.
45. Kutela B, Novat N, Langa N. Exploring geographical distribution of transportation research themes related to COVID-19 using text network approach. Sustain Cities Soc. 2021;67:102729.
46. Ashokkumar S, Kaushik NK, Han I, et al. Persistence of coronavirus on surface materials and its control measures using nonthermal plasma and other agents. Int J Mol Sci. 2023;24(18):14106.
47. Kwiatkowska A, Granicka LH. Anti-viral surfaces in the fight against the spread of coronaviruses. Membranes (Basel). 2023;13(5):464.
48. Wang H, Liu Q, Hu J, et al. Nasopharyngeal swabs are more sensitive than oropharyngeal swabs for COVID-19 diagnosis and monitoring the SARS-CoV-2 load. Front Med (Lausanne). 2020;7:334.
49. Lai CKC, Lam W. Laboratory testing for the diagnosis of COVID-19. Biochem Biophys Res Commun. 2021;538:226–30.
50. Corman VM, Landt O, Kaiser M, et al. Detection of 2019 novel coronavirus (2019-nCoV) by real-time RT-PCR. Eur Secur. 2020;25:2000045.
51. Tang YN, Jiang D, Wang X, et al. Recent progress on rapid diagnosis of COVID-19 by point-of-care testing platforms. Chin Chem Lett. 2023;2023:108688.
52. Khandker SS, Nik Hashim NHH, Deris ZZ, et al. Diagnostic accuracy of rapid antigen test kits for detecting SARS-CoV-2: a systematic review and meta-analysis of 17,171 suspected COVID-19 patients. J Clin Med. 2021;10(16):3493.
53. Murillo-Zamora E, Trujillo X, Huerta M, et al. Symptomatic SARS-COV-2 reinfection: healthcare workers and immunosuppressed individuals at high risk. BMC Infect Dis. 2021;21:923.
54. Scohy A, Anantharajah A, Bodéus M, et al. Low performance of rapid antigen detection test as front-line testing for COVID-19 diagnosis. J Clin Virol. 2020;129:104455.
55. Filip R, Gheorghita Puscaselu R, Anchidin-Norocel L, et al. Global challenges to public health care systems during the COVID-19 pandemic: a review of pandemic measures and problems. J Pers Med. 2022;12(8):1295.
56. Naseer S, Khalid S, Parveen S, et al. COVID-19 outbreak: impact on global economy. Front Public Health. 2022;10:1009393.
57. Adhikari SP, Meng S, Wu YJ, et al. Epidemiology, causes, clinical manifestation and diagnosis, prevention and control of coronavirus disease (COVID-19) during the early outbreak period: a scoping review. Infect Dis Poverty. 2020;9:29.
58. Komarova NL, Schang LM, Wodarz D. Patterns of the COVID19 pandemic spread around the world: exponential vs power laws. medRxiv 20202020033020047274.
59. Kamel Boulos MN, Geraghty EM. Geographical tracking and mapping of coronavirus disease COVID-19/severe acute respiratory syndrome coronavirus 2 (SARS-CoV-2) epidemic and associated events around the world: how 21st century GIS technologies are supporting the global fight against outbreaks and epidemics. Int J Health Geogr. 2020;19:8.

Epidemiology and Impact of the COVID-19 Pandemic

2

Samrah Rehan, Saurabh Mishra, Poorna Shri, Vishal Chand, Ashish Srivastava, and Supriya Raghav

2.1 Introduction

Throughout human history, microbial diseases—caused by bacteria, viruses, parasites, and fungi—have caused catastrophe [1]. Multiple drug-resistant (MDR), pan drug-resistant (PDR), and extensively drug-resistant (XDR) microbial infections, as well as re-emerging microbial diseases, are currently occurring at an accelerated rate [2–4]. According to the World Health Organization (WHO), throughout the past 10 years, more than 20 infectious agents have caused epidemics and a significant increase in infectious disease outbreaks on our planet [5]. These outbreaks were frequently caused by new viruses, including (MERS) Middle East respiratory syndrome and (H1N1), the novel influenza A virus [6–10].

The past two decades have seen the emergence of coronavirus-associated illnesses, severe acute respiratory syndrome (SARS), which have created significant global problems for their management and posed a threat to public health everywhere [11, 12]. The disease COVID-19 is triggered by infection with the SARS-CoV-2 virus, which is the most recent version of the β-coronavirus family. It is a 50–200 nm-diameter enveloped virus that contains single-stranded RNA [13–16]. On January 30, 2020, the World Health Organization (WHO) declared COVID-19 a global public health emergency, and subsequently characterized it as a pandemic on March 11, 2020 [17–19]. The World Health Organization's director general described COVID-19 as a persistent and well-established health problem on May 5, 2023, and no longer considered a Public Health Emergency of International Concern [20].

Authors "Samrah Rehan", "Saurabh Mishra", and "Poorna Shri" have equally contributed to this chapter.

S. Rehan (✉) · P. Shri
Department of Biochemistry, GSVM Medical College, Kanpur, India

S. Mishra · V. Chand
School of Life Sciences and Biotechnology, CSJM University, Kanpur, India
e-mail: vishalchand@csjmu.ac.in

A. Srivastava
Department of Zoology, Siddarth University, Siddharth Nagar, Uttar Pradesh, India
e-mail: ashishsri@suksn.edu.in

S. Raghav
School of Public Health, Poornima University, Jaipur, India

A. Raghav et al. (eds.), *COVID-19 Impact on Economy, Environment and Healthcare*,
https://doi.org/10.1007/978-981-95-1196-9_2

2.2 Understanding the Coronavirus

SARS-CoV-2 is a novel coronavirus, a positive-sense single-stranded RNA virus, that causes coronavirus disease 2019 (COVID-19) [21]. There are four different types of coronaviruses: Alpha (α), Beta (β), Gamma (γ), and Delta (δ) (Fig. 2.1). Although the Gamma and Delta subtypes primarily infect birds, the Alpha and Beta subtypes have clinical significance because they can infect people and cause infectious illnesses [22]. Alpha and Beta subtypes are associated with bats, while Gamma and Delta subtypes originated in birds [23].

In the middle of the 1960s, the first human coronavirus was discovered. There are now seven different kinds of coronaviruses that can infect people. The Alpha coronaviruses 229E and NL63, as well as the Beta coronaviruses OC43 and HKU1, are common human coronaviruses [24–26]. Mild upper respiratory tract infections can be brought on by these viruses [26]. Other human coronavirus strains, nevertheless, have the potential to cause serious respiratory infections. They are beta coronaviruses, specifically MERS-CoV, SARS-CoV, and SARS-CoV-2 [27, 28].

2.3 The Origin of COVID-19: How the Infection Began

In December 2019, hospitals in Wuhan, Hubei Province, China, reported several unexplainable pneumonia cases, marking the first COVID-19 outbreak. After the Huanan Wet Seafood Wholesale Market in Wuhan reported the first occurrence, it began to spread throughout the world. The initial suspicion was that the cause was viral pneumonia [29].

Lower respiratory tract samples from these patients were analyzed with unbiased next-generation sequencing, leading to the discovery of an unidentified beta coronavirus. Epithelial cells lining the respiratory tract were then used to isolate the 2019 novel coronavirus [30]. Under an electron microscope, the virus resembled the Coronoviridae family, and its diameter ranged from 60 to 140 nm, with distinctive spikes measuring 9–12 nm [30]. The novel coronavirus was discovered to have a higher phylogenetic resemblance (~88%) to two coronavirus strains derived from bats than to human-infecting coronaviruses, like SARS (around 79%) and MERS (around 50%) [31].

On February 11, 2020, the Coronaviridae Study Group of the International Committee on

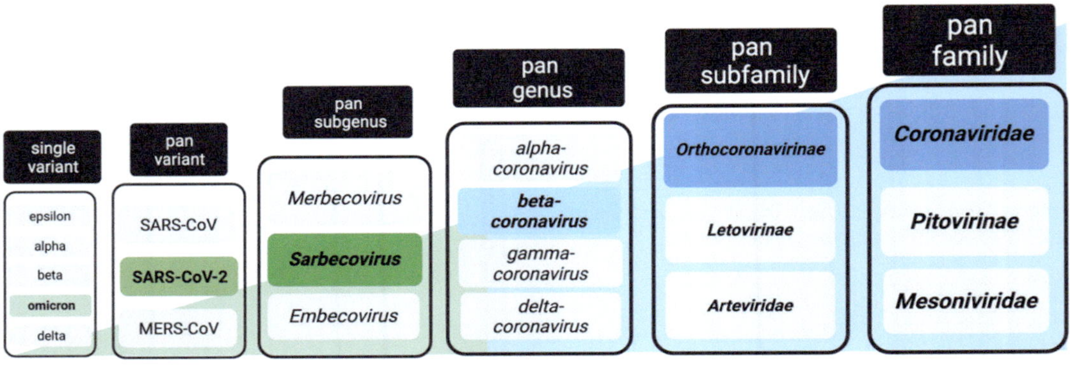

Fig. 2.1 Phylogeny of coronaviruses. (Adopted from Reference No. 22 under the terms and conditions of common creative license 4.0)

Taxonomy of Viruses study group designated the virus as SARS–CoV2 based on taxonomy and phylogeny [13]. The resulting illness was dubbed coronavirus disease (COVID-19) by the World Health Organization (WHO) [32]. After evaluating the global scenario, WHO classified COVID-19 as a pandemic on March 11, 2020.

At first, zoonotic transmission appeared to be a probable cause since it was typical for the first few cases to have experienced being around wet marketplaces [33]. Toward the end of January 2020, however, more persons had contracted the illness despite not being in contact with the market or another individual exhibiting respiratory symptoms. The disease's proliferation among healthcare personnel and those who did not had a travel history to Wuhan indicated that the virus transmitted from person to person [34, 35].

Severe COVID-19 individuals were thought to be more contagious than mild ones, and they were the primary source of infection. Potential sources of infection may also include asymptomatic individuals or incubating patients who exhibit no respiratory illness symptoms and have been shown to sweat infectious viruses [36]. Furthermore, samples obtained from COVID-19 patients consistently exhibit a positive RT-PCR test [37], which had never been observed throughout the timeline of infectious diseases in humans. Stated differently, individuals who are asymptomatic and patients who are incubating or have recovered from COVID-19 may provide significant obstacles to the prevention and management of the disease.

2.4 The Spectrum of Infection: Clinical Features and Disease Progression

Most COVID-19 individuals with moderate symptoms can recover in 1–2 weeks, making it a self-limiting infectious condition. There are five possible outcomes from SARS-CoV-2 infection: mild to medium cases (80.9%), critical cases (4.7%), severe cases (13.8%), asymptomatic individuals (1.2%), and death (2.3% in all recorded cases) [38]. Up to 15.8% of children under the age of ten have an asymptomatic infection [39] (Table 2.1). Thus, further research should be done in the future to determine the percentage of asymptomatic infections.

Table 2.1 Clinical spectrum of COVID-19 infection

Outcome	Percentage of cases (%)	Key characteristics
Mild to moderate	80.9	Self-limiting; recovery in 1–2 weeks.
Severe	13.8	Requires hospitalization; may develop acute pneumonia.
Critical	4.7	ICU admission; risk of cytokine storm and multi-organ failure.
Asymptomatic	1.2	No symptoms; higher prevalence in children (15.8% under age 10).
Fatal	2.3	Highest risk among elderly and comorbid patients (e.g., diabetes, cardiovascular disease).

Among the initial 41 patients [40], fever was the most prevalent symptom, occurring in 98% of cases, followed by cough at 76%, and myalgia or fatigue in 44%. Conversely, less common symptoms included diarrhea (3%), hemoptysis (5%), headache (8%), and sputum production (28%). More than half of the patients had dyspnea. Estimates showed that the average incubation period was 5.2 days, and the basic reproduction number (R0) was 2.2. The results of blood tests indicated that 25% of patients had either normal or decreased white blood cell counts, and 65% exhibited lymphopenia [40].

Under chest CT, bilateral involvement was seen in 98% of patients. When ICU patients were admitted, their chest CT scans commonly revealed multiple areas of consolidation in both lungs, affecting the lobes and subsegments. In non-ICU patients, representative chest CT findings revealed bilateral ground-glass opacity and consolidation in subsegmental areas [30, 40]. While diarrhea is rare, fever (87.9%) as well as cough (67.7%) remain the most common symptoms, according to an analysis of 1324 cases that were laboratory-confirmed. 82.1% of patients in the ICU presented with lymphopenia [41].

2.5 Understanding the Pathways of Disease Transmission

The three main categories of SARS-CoV-2 transmission are direct contact or indirect contact (via fomites), small airborne droplets (aerosols), and large droplets.

Direct or indirect contact is the initial mode of transmission. Both the upper and lower respiratory tracts are impacted by SARS-CoV-2. When it comes to direct contact, SARS-CoV-2 can be spread by meeting a COVID-19 patient's mouth, ears, or ocular mucous membranes [42]. Healthcare providers commonly contracted hospital-acquired infections through this transmission method during the COVID-19 pandemic [43]. Because they had to care for those with COVID-19 throughout the pandemic, healthcare professionals were particularly vulnerable to this kind of transmission [44]. SARS-CoV-2 is transmitted by fomites carrying infectious virus particles via indirect contact [45]. The contagious virus particles transmit from an infected individual and can adhere to any inanimate object's surface. Another person may become infected with SARS-CoV-2 if they encounter contaminated surface and transfer the infectious virus particles to their mouth, ears, or eyes' mucous membranes [46].

Aerosol transmission, the second mode, involves the spread of pathogens via microscopic airborne particles. Infection occurs when these particles enter the mucous membranes of the mouth, ears, or eyes. In enclosed spaces, aerosols can remain suspended for extended periods [47]; SARS-CoV-2, for instance, can survive in aerosols for at least 3 h [48]. Ventilation and air filtration can mitigate long-range aerosol transmission. Symptomatic patients release these infectious aerosols through coughing and sneezing [49]. SARS-CoV-2 can also be disseminated by asymptomatic COVID-19 patients who produce aerosols containing infectious SARS particles [50, 51]. They may release a lot of aerosols if they engage in exhalatory activities like talking, singing, or breathing [52]. If ventilation is absent, these pathogenic aerosols can linger in the atmosphere for a considerable amount of time.

Large droplets are the third method of transmission. If a COVID-19-infected individual coughs or sneezes, he may create large droplets that contain SARS CoV-2 virions. Should these big droplets be swallowed or inhaled, they can infect any vulnerable host [44]. SARS-CoV-2 can be prevented from spreading via droplets by keeping a minimum of 1.5 m away from an infected patient since huge droplets are only transmitted across short distances [53].

2.6 Pathogenesis and Infection Stages of SARS-CoV-2

SARS-CoV-2 comprises four structural proteins: spike, envelope, membrane, and nucleocapsid [54]. The spike protein contains two functional domains: S1 and S2 [55]. The S1 domain is responsible for receptor binding, while the S2 domain facilitates viral membrane fusion with the host cell [56]. The N-terminal domain (NTD) and receptor-binding domain (RBD) are two further divisions of the S1 domain. Cell receptors interact with the RBD, which is the most changeable component [57]. Like SARS-CoV, SARS-CoV-2 enters host cells through angiotensin-converting enzyme 2 (ACE2) receptors [58]. The proteolytic processing of S proteins, which facilitates membrane fusion and entry, depends on transmembrane serine protease 2 (TMPRSS2) [59]. The kidney, heart, lungs, and intestines all contain large amounts of ACE2 [60]. Depending on where ACE2 binding occurs, the spike protein's attachment to the ACE2 receptor initiates SARS-CoV-2 infection and may give rise to a wide array of symptoms [61]. For example, binding to ACE2 in the heart can cause myocarditis, while binding to ACE2 in the lungs can cause lung inflammation [62, 63].

Inflammation is part of the innate immune response to SARS-CoV-2. Upon entering cells, SARS-CoV-2 activates pattern recognition receptors (PRRs) to recognize pathogen-associated molecular patterns (PAMPs) and initiate an innate antiviral response [64]. Proinflammatory cytokines (IL-6) and chemokines (CCL2) are produced by host immune cells in response to the virus [65]. When PRRs recognize PAMP, they also release type I interferons (IFN-1), which cause antiviral responses [66, 67]. IFN-1 can trigger the production of enzymes that break down viral nucleic acids, as well as proteins that stop viral particles from being released from infected cells [67]. However, by precluding the manufacture of IFN-1, SARS-CoV-2 can reverse this. Viral non-structural proteins could degrade IFN-1 mRNA and inhibit the signaling process that leads to IFN-1 production [68].

A cytokine storm frequently occurs in COVID-19 patients with critical illness which is a poor prognostic indicator and can result from excessive proinflammatory cytokine release, such as IL-6 [69]. ACE2 receptors on macrophages and the viral spike protein can interact to cause these storms [70]. The cytokine storm can lead to tissue damage in various organs, including the heart, liver, and kidneys, potentially resulting in organ failure [71].

In reaction to SARS-CoV-2, the body's immune system produces antibodies and kills the virus through CD8+ lymphocytes. To produce IgM, IgG, and IgA antibodies against SARS-CoV-2, CD4+ cells must coordinate the humoral response [72]. By binding to the viral spike protein or RBD, neutralizing antibodies can stop the infection of host cells [73]. Additionally, cellular phagocytosis and antibody-dependent cell-mediated cytotoxic effects are facilitated by antibodies [74].

Individuals infected with SARS-CoV-2 typically experience three stages. The initial phase is known as the incubation period, during which the individual contracts SARS-CoV-2. Viruses begin to replicate at this point, and symptoms typically do not appear. The second stage, known as the dissemination stage, is when patients may experience acute pneumonia in addition to other symptoms like fever and cough [75]. Hypoxia is one of the more severe symptoms that some individuals with diabetes or cardiovascular disease may encounter [76]. The most severe level, stage three, will occur if the patient fails to recover from the second stage. Stage three patients may experience systemic hyperinflammation, which can lead to gastrointestinal, hepatocellular, and renal damage [77, 78]. Comorbidity is one of the leading reasons of death for COVID-19 patients. A patient is said to be comorbid if they have a medical ailment in addition to their initial disease diagnosis. Heart disease, kidney disease, diabetes, obesity, and hypertension are examples of common comorbidities. One important risk factor that can raise the likelihood of death is comorbidity [79]. The probability of contracting COVID-19 and the severity of the disease in patients with comorbidities increased as ACE2 expression increased [79]. Additionally, some people are multimorbid, meaning they have multiple comorbidities. Patients with multimorbidity had a twice as high mortality risk as those without multimorbidity [80].

2.7 Current Scenario of the COVID-19 Pandemic

As of April 6, 2025, altogether, 777,704,325 COVID-19 cases have been documented globally, reflecting an increase of 6435 cases over the previous 7 days. The United States has reported the record number of cases at approximately 103 million, followed by China with 99.4 million and India with 45 million [81]. The global death toll due to COVID-19 stands at 7,093,786, with 331 new deaths reported in the past week. The United States has recorded the largest number of fatalities caused by COVID-19 with 1.2 million, followed by Brazil with 703,000 and India with 534,000 [82] (Table 2.2).

The epidemiology for the 4 weeks of January 6–February 2, 2025, has been updated by WHO and can be summed up as follows:

- Systematic virological surveillance showed that weekly SARS-CoV-2 PCR positivity increased from 7.3% in the first week of the reporting period to 5.0% in the last week. During the 4-week reporting period (January 6–February 2, 2025), an average of over 69,900 specimens were tested weekly across

Table 2.2 Global COVID-19 cases and deaths (as of April 6, 2025)

Metric	Global total	United States	China	India
Total cases	777,704,325	103 million	99.4 million	45 million
Total deaths	7,093,786	1.2 million	Not specified	534,000
New cases (past 7 days)	6435	Not specified	Not specified	Not specified
New deaths (past 7 days)	331	Not specified	Not specified	Not specified

Table 2.3 SARS-CoV-2 variants under surveillance (January–February 2025)

Variant category	Variant name	Prevalence (Week 5, 2025) (%)	Trend
Variant of Interest (VOI)	JN.1	16.3	Increasing
Variant Under Monitoring (VUM)	XEC	42.7	Declining
Variant Under Monitoring (VUM)	LP.8.1	13.9	Increasing
Variant Under Monitoring (VUM)	LB.1	1.2	Increasing

103 countries. No region reported a higher percentage of positive tests in the final week.

- WHO is keeping an eye on eight SARS-CoV-2 variants, including seven variants under monitoring (VUMs) and one variant of interest (VOI) JN.1. In week 5 (ending February 2, 2025), 16.3% of sequences were JN.1, the VOI. The most common VUM XEC, which accounted for 42.7% of sequences, demonstrated a decline in prevalence. As of week 5 of 2025, the only tracked variants that are actively increasing in prevalence are LP.8.1 and LB.1, which accounted for 13.9% and 1.2% of sequences, respectively. The prevalence of each of the remaining VUMs is decreasing (Table 2.3).
- One crucial aspect of SARS-CoV-2 surveillance is wastewater surveillance, which is also used for early warning and to track the spread of SARS-CoV-2 variants. Wastewater surveillance data from five WHO Regions and about 30 countries are publicly accessible and shown on WHO's COVID-19 dashboard. The SARS-CoV-2 virus circulates between two and nineteen times more frequently than the number of cases that are detected and reported, according to estimates derived from wastewater surveillance.
- Between January 6 and February 2, 2025, 83 countries (35%) reported COVID-19 cases, and 23 countries (10%) reported deaths. Compared to the preceding 28-day period (December 9, 2024, to January 5, 2025), reported cases decreased by 16% (with over 147,000 new cases), while new deaths increased by 28% (with over 4500 fatalities). It is important to interpret these trends in new cases and deaths cautiously, due to reduced testing and sequencing, and reporting delays in many countries.
- Of the nations that submitted data over the same period, 33 (14%) reported at least one ICU admission, while 43 (18%) reported COVID-19 hospitalizations. More than 700 additional intensive care unit admissions and more than 16,700 new hospitalizations were reported during this time, according to the data that is currently available. Compared to the preceding 28 days, there was an overall decrease of 40% in new hospitalizations and 31% in ICU admissions among the nations that consistently reported these data during the current and prior reporting periods.
- For health systems, post-COVID-19 condition (PCC) remains a significant burden. The incidence of PCC is difficult to pinpoint with great accuracy, although statistics indicate that PCC symptoms occurred in about 6% of symptomatic SARS-CoV-2 infections. Even though severe COVID-19 is a major risk factor for PCC, the sheer number of infections causes over 90% of PCC cases to occur after

mild COVID-19. There seems to be a preventive effect of vaccination, which lowers the risk of PCC [83].

2.8 The Global Impact of COVID-19

The COVID-19 pandemic has profoundly impacted daily life. The pandemic lockdown, intended to curb virus transmission and slow the epidemic's spread, significantly affected various aspects of society, including food security, the global economy, education, tourism, hospitality, sports, leisure, gender relations, domestic violence/abuse, mental health, and environmental air pollution. This lockdown has had extensive consequences, including changes in education delivery, food insecurity due to price volatility and scarcity, a global economic downturn, and increased mental health challenges, impacting overall wellbeing and quality of life [84].

2.8.1 The Economic Impact of the Global Lockdown

It should not be shocking that the lockdown is influencing the world economy. International trade, finance, and investments have changed and fluctuated because of many nations' restrictive policies. The COVID-19 pandemic and resulting restrictions caused international trade between the global South and North to fluctuate [85]. In the UK, lockdowns in the second quarter of 2020 led to a decrease in both imports and exports. However, as restrictions eased in the third quarter, the trade of goods saw an increase [86].

At the start of the lockdown in March 2020, major indices around the world fell, further demonstrating the impact of the shutdown on the financial market [87]. Several of the largest companies in the world had a sharp decline in share values during this time. The ability of many African nations to conduct commerce was impacted by the dollar's sharp increase in value relative to most other currencies. The global stock market experienced a sharp downturn on March 23, with its value dropping by a massive $26 trillion. This steep decline resulted in substantial financial losses for both individual shareholders and institutions holding investments in stocks, such as insurance companies and pension funds [88] (Table 2.4).

Some countries chose to delay implementing lockdowns primarily due to concerns about the economy, even though earlier action might have ultimately reduced the overall economic impact and GDP losses. This decision to prioritize economic factors over immediate health concerns led to a higher number of COVID-19 infections [89]. This is seen in the instance of the United Kingdom, where expert advice and communications emphasized the advantages of an early lockdown in slowing the spread of the virus, but the execution of the lockdown was postponed for 2 weeks [90]. In the same way, several countries began lifting their lockdown measures prematurely, disregarding the World Health Organization's recommended strategies for planning, preparedness, and response. This was largely driven by the economic strain and financial difficulties caused by the prolonged lockdowns.

2.8.2 How the Global Lockdown Disrupted Education Systems

Globally, most governments implemented widespread school closures at all levels as a key

Table 2.4 Economic impact of COVID-19 lockdowns

Sector	Impact	Estimated loss
Global Stock Market	Drop of $26 trillion (March 2020).	$26 trillion (March 23, 2020)
International Trade	UK: 22% reduction in Q1 2020; global exports fell by 60–80% in 2020.	$80 billion (tourism revenue lost)
Tourism	67 million fewer international visitors (Q1 2020).	$80 billion (export revenue lost)
Hospitality	Event cancellations, hotel closures, Olympics postponed to 2021.	Not quantified

measure to control the transmission of the virus during lockdowns. With 143 countries implementing nationwide school closures, this has had an impact on the formal education sector. Many universities throughout the world have quickly adopted online learning to conduct lectures and other academic tasks [91]. The shift to online education presents significant challenges in developing countries due to poor internet infrastructure, unaffordable internet access, and a shortage of devices available for students. This is made worse by a lack of electricity and teachers who are not proficient in using digital technology in the classroom.

Additionally, because their education is disrupted, young people are affected in a variety of ways by school closures, school year repeat, or even failing national exams. Additionally, working on the farm or performing housework had caused a lack of drive and focus for studying [92].

In addition to the impact that school closures had on students and youth, the lockdown has a variety of repercussions on scholars and research across different educational institutions. Because fieldwork had been shut off and laboratory access had been restricted, some researchers had been unable to conduct any study at all because of the lockdown. This also implies that new technology had been adopted and new methods of working have been learned. For others, the lockdown has resulted in less administrative work and shorter meetings, which causes tension in the classroom and on the mind. Because there is more time to ponder, reflect, and concentrate, research productivity has increased as a result [93].

2.8.3 The Economic Toll of Global Lockdowns on Tourism and Hospitality

The global lockdown caused a significant contraction in international travel. It saw a 22% reduction in the first 3 months of 2020, and forecasts suggested an even more substantial decrease, between 60% and 80%, by the close of that year. Overall, from the beginning of Q1 2020 until the end of March, there were 67 million fewer foreign visitors, which is estimated to represent a $80 billion loss in missed export revenue.

Notably, nations that are considered COVID-19 hotspots have lost millions of dollars in revenue because of people cancelling plans to visit vacation destinations due to the lockdown. Global tourism had also slowed down considerably due to the restrictions on travel and the implementation of lockdowns globally; international flights declined over time, and an increasing number of carriers had been grounded due to travel bans [94].

With event cancellations, hotel and lodging closures, leisure park closures, and restaurant meals and services, the hospitality value chain was also impacted. Just like the tourism sector, numerous recreational facilities such as parks, gyms, and swimming pools were shut down. Sporting events across the globe were either cancelled or attended by no spectators worldwide. The Olympic Games in Tokyo in 2020 are a prominent example. The event had been rescheduled for 2021 [95]. This situation arose because these events typically involve large gatherings and proximity among people, which significantly elevates the risk of the virus spreading [96].

2.8.4 The Mental Health Impact of the Global Lockdown

While crucial for controlling the spread of the virus, the lockdown measures put in place may have taken a considerable toll on people's mental well-being [97]. Post-traumatic stress disorder, anxiety, sadness, discomfort, and sleep difficulties are a few examples. The length of the lockdown, fear of infections, boredom, and frustration, lacking supplies, and lack of information are the five primary reasons for psychological suffering during the lockdown [98]. Anxiety and despair levels are higher during pandemics like this one than they were before, according to reports [99].

Several factors linked to the COVID-19 pandemic, such as the fear of contracting the virus or dying, contributed to a rise in mental health issues. Consequently, individuals who survived the illness reported increased feelings of confusion, anger, worry, and symptoms like those experienced after a traumatic event [100, 101]. Feelings of helplessness, sadness, frustration, and loneliness have been commonly reported, and these emotions can be intensified by isolating oneself or being quarantined, the monotony of limited routines, the spread of misinformation, and financial stress. Tragically, these combined pressures could increase the risk of harmful actions, including self-harm, suicidal ideation, or even suicide [102, 103].

2.9 COVID-19: Lessons Learned and the Future Directions

The world was caught off guard by the COVID-19 epidemic, which disrupted daily life worldwide and exposed serious weaknesses in pandemic preparedness. In response, governments, public health organizations, scientists, and communities came together in a never-before-seen worldwide endeavor to control the virus, put safety measures in place, and create vaccines, treatments, and diagnostic tools. Despite the dangers posed by a lack of personal protective equipment (PPE), medical personnel continued to serve on the front lines, and the World Health Organization (WHO) organized global initiatives to provide evidence-based recommendations.

Even though the WHO no longer considers COVID-19 a global health emergency, the fact that the SARS-CoV-2 virus keeps changing, raising the risk of new versions that vaccines might not protect against, means it is still a serious worldwide health issue. The global network of scientists has improved our knowledge of the virus's intricate dynamics since the beginning of the pandemic, which has influenced the creation of vaccines and therapeutic approaches. The possibility of novel, more contagious, or immune-evasive variations, however, emphasizes the necessity of ongoing monitoring, situational awareness, and preparedness for future waves.

The management of COVID-19 as an endemic illness is currently the main emphasis, with special attention paid to seasonality and co-circulation with other respiratory viruses like influenza and RSV. Healthcare systems may be overloaded by concurrent spikes in these infections if proper monitoring and response measures are not in place. To improve vaccination coverage and population protection overall, the development of a combined COVID-19/influenza vaccine and improved public education may assist in fighting vaccine hesitancy and booster weariness.

2.10 Conclusion

The epidemic has taught us important lessons about international cooperation, quick scientific advancement, and public health adaptability. Even though COVID-19 has been significantly reduced, continued research and funding for the creation of vaccines and treatments are necessary to keep the virus under control over the long run and get ready for any future health emergencies. To prevent and lessen the impact of future pandemics, a more proactive, methodical, and internationally coordinated strategy will be essential.

References

1. Balkhair AA. COVID-19 pandemic: a new chapter in the history of infectious diseases. Oman Med J. 2020;35(2):e123.
2. Basak S, Singh P, Rajurkar M. Multidrug resistant and extensively drug resistant bacteria: a study. J Pathog. 2016;2016(1):4065603.
3. Magiorakos AP, Srinivasan A, Carey RB, Carmeli Y, Falagas ME, Giske CG, et al. Multidrug-resistant, extensively drug-resistant and pan drug-resistant bacteria: an international expert proposal for interim standard definitions for acquired resistance. Clin Microbiol Infect. 2012;18(3):268–81. https://doi.org/10.1111/j.1469-0691.2011.03570.x.
4. Morens DM, Fauci AS. Emerging infectious diseases: threats to human health and global stability. PLoS Pathog. 2013;9(7):e1003467.
5. Haque M. The COVID-19 pandemic—a global public health crisis: a brief overview regarding pharmacological interventions. Pesqui Bras Odontopediatria Clin Integr. 2020;20:e0146.
6. Balkhair A. The struggle against pandemic influenza A (H1N1) 2009. Sultan Qaboos Univ Med

J. 2009;9(3):257–60. Epub 2009 Dec 19. PMID: 21509307; PMCID: PMC3074803.
7. Triana-Baltzer GB, Gubareva LV, Nicholls JM, Pearce MB, Mishin VP, Belser JA, Chen LM, Chan RW, Chan MC, Hedlund M, Larson JL. Novel pandemic influenza A (H1N1) viruses are potently inhibited by DAS181, a sialidase fusion protein. PLoS One. 2009;4(11):e7788.
8. Balkhair A, Al Maamari K, Alawi FB. The struggle against MERS-CoV (the novel coronavirus). Oman Med J. 2013;28(4):226.
9. Triana-Baltzer GB, Gubareva LV, Klimov AI, Wurtman DF, Moss RB, Hedlund M, Larson JL, Belshe RB, Fang F. Inhibition of neuraminidase inhibitor-resistant influenza virus by DAS181, a novel sialidase fusion protein. PLoS One. 2009;4(11):e7838.
10. Sutton TC, Subbarao K. Development of animal models against emerging coronaviruses: from SARS to MERS coronavirus. Virology. 2015;479:247–58.
11. De Wit E, Van Doremalen N, Falzarano D, Munster VJ. SARS and MERS: recent insights into emerging coronaviruses. Nat Rev Microbiol. 2016;14(8):523–34.
12. Liu J, Zheng X, Tong Q, Li W, Wang B, Sutter K, Trilling M, Lu M, Dittmer U, Yang D. Overlapping and discrete aspects of the pathology and pathogenesis of the emerging human pathogenic coronaviruses SARS-CoV, MERS-CoV, and 2019-nCoV. J Med Virol. 2020;92(5):491–4.
13. Coronaviridae Study Group of the International Committee on Taxonomy of Viruses. The species severe acute respiratory syndrome-related coronavirus: classifying 2019-nCoV and naming it SARS-CoV-2. Nat Microbiol. 2020;5(4):536–44.
14. Wenbin YU, Guangda TANG, Li ZHANG. Analysis of the evolution and spread of novel coronavirus based on genome-wide data. Zool Res. 2020;41(3):247–57.
15. Malik YS, Sircar S, Bhat S, Sharun K, Dhama K, Dadar M, Tiwari R, Chaicumpa W. Emerging novel coronavirus (2019-nCoV)—current scenario, evolutionary perspective based on genome analysis and recent developments. Vet Q. 2020;40(1):68–76.
16. Xu X, Chen P, Wang J, Feng J, Zhou H, Li X, Zhong W, Hao P. Evolution of the novel coronavirus from the ongoing Wuhan outbreak and modeling of its spike protein for risk of human transmission. Sci China Life Sci. 2020;63:457–60.
17. Gallegos A. WHO declares public health emergency for novel coronavirus. Medscape Med News. 2020;30(1). https://www.medscape.com/viewarticle/924596
18. Ramzy A, McNeil DG. WHO declares global emergency as Wuhan coronavirus spreads. The New York Times; 2020.
19. Castle S. Coronavirus live updates: WHO declares pandemic as number of infected countries grows. The-New-York-Times; 2020.
20. World Health Organization. Statement on the fifteenth meeting of the IHR (2005) Emergency Committee on the COVID-19 pandemic. [Online]. World Health Organization. 2023. Available online: https://www.who.int/news/item/05-05-2023-statement-on-the-fifteenth-meeting-of-the-international-health-regulations-(2005)-emergency-committee-regarding-the coronavirus-disease-(covid-19)-pandemic
21. Tsang HF, Chan LW, Cho WC, Yu AC, Yim AK, Chan AK, Ng LP, Wong YK, Pei XM, Li MJ, Wong SC. An update on COVID-19 pandemic: the epidemiology, pathogenesis, prevention, and treatment strategies. Expert Rev Anti-Infect Ther. 2021;19(7):877–88.
22. Cankat S, Demael MU, Swadling L. In search of a pan-coronavirus vaccine: next-generation vaccine design and immune mechanisms. Cell Mol Immunol. 2024;21(2):103–18. https://doi.org/10.1038/s41423-023-01116-8.
23. Woo PC, Lau SK, Huang Y, Yuen KY. Coronavirus diversity, phylogeny, and interspecies jumping. Exp Biol Med. 2009;234(10):1117–27.
24. Myint SH. Human coronavirus infections. In: The coronaviridae. New York: Plenum Press; 1995. p. 389–401.
25. Vabret A, Mourez T, Dina J, Van Der Hoek L, Gouarin S, Petitjean J, Brouard J, Freymuth F. Human coronavirus NL63, France. Emerg Infect Dis. 2005;11(8):1225.
26. Woo PC, Lau SK, Chu CM, Chan KH, Tsoi HW, Huang Y, Wong BH, Poon RW, Cai JJ, Luk WK, Poon LL. Characterization and complete genome sequence of a novel coronavirus, coronavirus HKU1, from patients with pneumonia. J Virol. 2005;79(2):884–95.
27. Ogimi C, Kim YJ, Martin ET, Huh HJ, Chiu CH, Englund JA. What is new with the old coronaviruses? J Pediatr Infect Dis Soc. 2020;9(2):210–7.
28. De Groot RJ, Baker SC, Baric RS, Brown CS, Drosten C, Enjuanes L, Fouchier RA, Galiano M, Gorbalenya AE, Memish ZA, Perlman S. Commentary: middle east respiratory syndrome coronavirus (mers-cov): announcement of the coronavirus study group. J Virol. 2013;87(14):7790–2.
29. Luo GG, Gao SJ. Global health concerns stirred by emerging viral infections. J Med Virol. 2020;92(4):399.
30. Zhu N, Zhang D, Wang W, Li X, Yang B, Song J, Zhao X, Huang B, Shi W, Lu R, Niu P. A novel coronavirus from patients with pneumonia in China, 2019. N Engl J Med. 2020;382(8):727–33.
31. Lu R, Zhao X, Li J, Niu P, Yang B, Wu H, Wang W, Song H, Huang B, Zhu N, Bi Y. Genomic characterization, and epidemiology of 2019 novel coronavirus: implications for virus origins and receptor binding. Lancet. 2020;395(10224):565–74.
32. World Health Organization. Naming the coronavirus disease (COVID-19) and the virus that causes it. Braz J Implantol Health Sci. 2020;2(3). https://www.who.

int/emergencies/diseases/novel-coronavirus-2019/technical-guidance/naming-the-coronavirus-disease-(covid-2019)-and-the-virus-that-causes-it
33. Li Q, Guan X, Wu P, Wang X, Zhou L, Tong Y, Ren R, Leung KS, Lau EH, Wong JY, Xing X. Early transmission dynamics in Wuhan, China, of novel coronavirus–infected pneumonia. N Engl J Med. 2020;382(13):1199–207.
34. Ralph R, Lew J, Zeng T, Francis M, Xue B, Roux M, Ostadgavahi AT, Rubino S, Dawe NJ, Al-Ahdal MN, Kelvin DJ. 2019-nCoV (Wuhan virus), a novel coronavirus: human-to-human transmission, travel-related cases, and vaccine readiness. J Infect Dev Ctries. 2020;14(01):3–17.
35. Chan JF, Yuan S, Kok KH, To KK, Chu H, Yang J, Xing F, Liu J, Yip CC, Poon RW, Tsoi HW. A familial cluster of pneumonia associated with the 2019 novel coronavirus indicating person-to-person transmission: a study of a family cluster. Lancet. 2020;395(10223):514–23.
36. Hoehl S, Rabenau H, Berger A, Kortenbusch M, Cinatl J, Bojkova D, Behrens P, Böddinghaus B, Götsch U, Naujoks F, Neumann P. Evidence of SARS-CoV-2 infection in returning travelers from Wuhan, China. N Engl J Med. 2020;382(13):1278–80.
37. Lan L, Xu D, Ye G, Xia C, Wang S, Li Y, Xu H. Positive RT-PCR test results in patients recovered from COVID-19. JAMA. 2020;323(15):1502–3.
38. Team E. The epidemiological characteristics of an outbreak of 2019 novel coronavirus diseases (COVID-19)—China, 2020. China CDC Weekly. 2020;2(8):113.
39. Lu X, Zhang L, Du H, Zhang J, Li YY, Qu J, Zhang W, Wang Y, Bao S, Li Y, Wu C. SARS-CoV-2 infection in children. N Engl J Med. 2020;382(17):1663–5.
40. Huang C, Wang Y, Li X, Ren L, Zhao J, Hu Y, Zhang L, Fan G, Xu J, Gu X, Cheng Z. Clinical features of patients infected with 2019 novel coronavirus in Wuhan, China. Lancet. 2020;395(10223):497–506.
41. Yang Y, Lu QB, Liu MJ, Wang YX, Zhang AR, Jalali N, Dean NE, Longini I, Halloran ME, Xu B, Zhang XA. Epidemiological and clinical features of the 2019 novel coronavirus outbreak in China. medrxiv. 2020;2020-02
42. Belser JA, Rota PA, Tumpey TM. Ocular tropism of respiratory viruses. Microbiol Mol Biol Rev. 2013;77(1):144–56.
43. Lu Q, Shi Y. Coronavirus disease (COVID-19) and neonate: what neonatologist need to know. J Med Virol. 2020;92(6):564–7.
44. Li H, Leong FY, Xu G, Kang CW, Lim KH, Tan BH, Loo CM. Airborne dispersion of droplets during coughing: a physical model of viral transmission. Sci Rep. 2021;11(1):4617.
45. Jiang F, Deng L, Zhang L, Cai Y, Cheung CW, Xia Z. Review of the clinical characteristics of coronavirus disease 2019 (COVID-19). J Gen Intern Med. 2020;35:1545–9.
46. Chen T. Fomites and the COVID-19 pandemic: an evidence review on its role in viral transmission. Vancouver: National Collaborating Centre for Environmental Health; 2021. p. 1–24.
47. Wu D, Wu T, Liu Q, Yang Z. The SARS-CoV-2 outbreak what we know. Int J Infect Dis. 2020;94:44–8.
48. Van Doremalen N, Bushmaker T, Morris DH, Holbrook MG, Gamble A, Williamson BN, Tamin A, Harcourt JL, Thornburg NJ, Gerber SI, Lloyd-Smith JO. Aerosol and surface stability of SARS-CoV-2 as compared with SARS-CoV-1. N Engl J Med. 2020;382(16):1564–7.
49. Koenig KL, Bey CK, McDonald EC. 2019-nCoV: the identify-isolate-inform (3I) tool applied to a novel emerging coronavirus. West J Emerg Med. 2020;21(2):184.
50. Tellier R. COVID-19: the case for aerosol transmission. Interface Focus. 2022;12(2):20210072.
51. Zhou C. Evaluating new evidence in the early dynamics of the novel coronavirus COVID-19 outbreak in Wuhan, China with real time domestic traffic and potential asymptomatic transmissions. MedRxiv. 2020;2020-02
52. Morawska LJ, Johnson GR, Ristovski ZD, Hargreaves M, Mengersen K, Corbett S, Chao CY, Li Y, Katoshevski D. Size distribution and sites of origin of droplets expelled from the human respiratory tract during expiratory activities. J Aerosol Sci. 2009;40(3):256–69.
53. Cortellessa G, Stabile L, Arpino F, Faleiros DE, Van Den Bos W, Morawska L, Buonanno G. Proximity risk assessment for SARS-CoV-2 infection. Sci Total Environ. 2021;794:148749.
54. Akkiz H. Implications of the novel mutations in the SARS-CoV-2 genome for transmission, disease severity, and the vaccine development. Front Med. 2021;8:636532.
55. Andersen KG, Rambaut A, Lipkin WI, Holmes EC, Garry RF. The proximal origin of SARS-CoV-2. Nat Med. 2020;26(4):450–2.
56. Jiang S, Hillyer C, Du L. Neutralizing antibodies against SARS-CoV-2 and other human coronaviruses. Trends Immunol. 2020;41(5):355–9.
57. V'kovski P, Kratzel A, Steiner S, Stalder H, Thiel V. Coronavirus biology and replication: implications for SARS-CoV-2. Nat Rev Microbiol. 2021;19(3):155–70.
58. Hoffmann M, Kleine-Weber H, Schroeder S, Krüger N, Herrler T, Erichsen S, Schiergens TS, Herrler G, Wu NH, Nitsche A, Müller MA. SARS-CoV-2 cell entry depends on ACE2 and TMPRSS2 and is blocked by a clinically proven protease inhibitor. Cell. 2020;181(2):271–80.
59. Fuentes-Prior P. Priming of SARS-CoV-2 S protein by several membrane-bound serine proteinases could explain enhanced viral infectivity and systemic COVID-19 infection. J Biol Chem. 2021;296:100135.
60. Wu J, Deng W, Li S, Yang X. Advances in research on ACE2 as a receptor for 2019-nCoV. Cell Mol Life Sci. 2021;78:531–44.

61. Hanff TC, Harhay MO, Brown TS, Cohen JB, Mohareb AM. Is there an association between COVID-19 mortality and the renin-angiotensin system? A call for epidemiologic investigations. Clin Infect Dis. 2020;71(15):870–4.
62. Shieh WJ, Hsiao CH, Paddock CD, Guarner J, Goldsmith CS, Tatti K, Packard M, Mueller L, Wu MZ, Rollin P, Su IJ. Immunohistochemical, in situ hybridization, and ultrastructural localization of SARS-associated coronavirus in lung of a fatal case of severe acute respiratory syndrome in Taiwan. Hum Pathol. 2005;36(3):303–9.
63. Chen C, Zhou Y, Wang DW. SARS-CoV-2: a potential novel etiology of fulminant myocarditis. Herz. 2020;45(3):230–2.
64. Li D, Wu M. Pattern recognition receptors in health and diseases. Signal Transduct Target Ther. 2021;6(1):291.
65. Ebbole DJ. Magnaporthe as a model for understanding host-pathogen interactions. Annu Rev Phytopathol. 2007;45(1):437–56.
66. Streicher F, Jouvenet N. Stimulation of innate immunity by host and viral RNAs. Trends Immunol. 2019;40(12):1134–48.
67. Ivashkiv LB, Donlin LT. Regulation of type I interferon responses. Nat Rev Immunol. 2014;14(1):36–49.
68. Deng X, Hackbart M, Mettelman RC, O'Brien A, Mielech AM, Yi G, Kao CC, Baker SC. Coronavirus nonstructural protein 15 mediates evasion of dsRNA sensors and limits apoptosis in macrophages. Proc Natl Acad Sci. 2017;114(21):E4251–60.
69. Chen X, Zhao B, Qu Y, Chen Y, Xiong J, Feng Y, Men D, Huang Q, Liu Y, Yang B, Ding J. Detectable serum severe acute respiratory syndrome coronavirus 2 viral load (RNAemia) is closely correlated with drastically elevated interleukin 6 level in critically ill patients with coronavirus disease 2019. Clin Infect Dis. 2020;71(8):1937–42.
70. Paul BD, Lemle MD, Komaroff AL, Snyder SH. Redox imbalance links COVID-19 and myalgic encephalomyelitis/chronic fatigue syndrome. Proc Natl Acad Sci. 2021;118(34):e2024358118.
71. Channappanavar R, Perlman S. Pathogenic human coronavirus infections causes and consequences of cytokine storm and immunopathology. In: Seminars in immunopathology, vol. 39. Berlin/Heidelberg: Springer; 2017. p. 529–39.
72. Grifoni A, Weiskopf D, Ramirez SI, Mateus J, Dan JM, Moderbacher CR, Rawlings SA, Sutherland A, Premkumar L, Jadi RS, Marrama D. Targets of T cell responses to SARS-CoV-2 coronavirus in humans with COVID-19 disease and unexposed individuals. Cell. 2020;181(7):1489–501.
73. Shi D, Weng T, Wu J, Dai C, Luo R, Chen K, Zhu M, Lu X, Cheng L, Chen Q, Liu F. Dynamic characteristic analysis of antibodies in patients with COVID-19: a 13-month study. Front Immunol. 2021;12:708184.
74. Zohar T, Alter G. Dissecting antibody-mediated protection against SARS-CoV-2. Nat Rev Immunol. 2020;20(7):392–4.
75. Vetter P, Eberhardt CS, Meyer B, Martinez Murillo PA, Torriani G, Pigny F, Lemeille S, Cordey S, Laubscher F, Vu DL, Calame A. Daily viral kinetics, and innate and adaptive immune response assessment in COVID-19: a case series. MSphere. 2020;5(6):10–128.
76. Jin JM, Bai P, He W, Wu F, Liu XF, Han DM, Liu S, Yang JK. Gender differences in patients with COVID-19: focus on severity and mortality. Front Public Health. 2020;8:545030.
77. Gupta A, Madhavan MV, Sehgal K, Nair N, Mahajan S, Sehrawat TS, Bikdeli B, Ahluwalia N, Ausiello JC, Wan EY, Freedberg DE. Extrapulmonary manifestations of COVID-19. Nat Med. 2020;26(7):1017–32.
78. Mehta P, McAuley DF, Brown M, Sanchez E, Tattersall RS, Manson JJ. COVID-19: consider cytokine storm syndromes and immunosuppression. Lancet. 2020;395(10229):1033–4.
79. Singh MK, Mobeen A, Chandra A, Joshi S, Ramachandran S. A meta-analysis of comorbidities in COVID-19: which diseases increase the susceptibility of SARS-CoV-2 infection? Comput Biol Med. 2021;130:104219.
80. Agrawal U, Azcoaga-Lorenzo A, Fagbamigbe AF, Vasileiou E, Henery P, Simpson CR, Stock SJ, Shah SA, Robertson C, Woolhouse M, Ritchie LD. Association between multimorbidity and mortality in a cohort of patients admitted to hospital with COVID-19 in Scotland. J R Soc Med. 2022;115(1):22–30.
81. COVID-19 cases | WHO COVID-19 dashboard as on 6 April 2025.
82. COVID-19 deaths | WHO COVID-19 dashboard as on 6 April 2025.
83. Coronavirus Disease (COVID-19) Situation Reports.
84. Onyeaka H, Anumudu CK, Al-Sharify ZT, Egele-Godswill E, Mbaegbu P. COVID-19 pandemic: a review of the global lockdown and its far-reaching effects. Sci Prog. 2021;104(2):00368504211019854.
85. United Nations. Impact of the covid-19 pandemic on trade and development. Transitioning to a new normal. Impact of the COVID-19 pandemic on trade and development: transitioning to a new normal | UN Trade and Development (UNCTAD).
86. Impacts of the coronavirus on UK trade—Office for National Statistics.
87. Fernandes N. Economic effects of coronavirus outbreak (COVID-19) on the world economy; 2020. https://doi.org/10.2139/ssrn.3557504.
88. Tooze A. How coronavirus almost brought down the global financial system. The Guardian; 2020. p. 14.
89. Balmford B, Annan JD, Hargreaves JC, Altoè M, Bateman IJ. Cross-country comparisons of Covid-19: policy, politics, and the price of life. Environ Resour Econ. 2020;76(4):525–51.

90. Iacobucci G. Covid-19: UK lockdown is "crucial" to saving lives, say doctors and scientists. BMJ. 2020;368:m1204.
91. Cambridge University to move all lectures online until 2021 | The Independent | The Independent.
92. Parkes J, Datzberger S, Howell C, Knight L, Kasidi J, Kiwanuka T, Nagawa R, Naker D, Devries K. Young people, inequality, and violence during the COVID-19 lockdown in Uganda. London: UCL Institute of Education; 2020.
93. Eaton H. Science in lockdown: the effects of COVID-19 on research and researchers. The Royal Society; 2020. p. 4.
94. Gössling S, Scott D, Hall CM. Pandemics, tourism, and global change: a rapid assessment of COVID-19. J Sustain Tour. 2020;29(1):1–20.
95. Drewes M, Daumann F, Follert F. Exploring the sports economic impact of COVID-19 on professional soccer. In: COVID-19 and the soccer world. Routledge; 2022. p. 130–42.
96. Rashid H, Haworth E, Shafi S, Memish ZA, Booy R. Pandemic influenza: mass gatherings and mass infection. Lancet Infect Dis. 2008;8(9):526–7.
97. Atalan A. Is the lockdown important to prevent the COVID-19 pandemic? Effects on psychology, environment, and economy-perspective. Ann Med Surg. 2020;56:38–42.
98. Brooks SK, Webster RK, Smith LE, Woodland L, Wessely S, Greenberg N, Rubin GJ. The psychological impact of quarantine and how to reduce it: rapid review of the evidence. Lancet. 2020;395(10227):912–20.
99. Huang Y, Zhao N. Generalized anxiety disorder, depressive symptoms and sleep quality during COVID-19 outbreak in China: a web-based cross-sectional survey. Psychiatry Res. 2020;288:112954.
100. Mukhtar S. Pakistanis' mental health during the COVID-19. Asian J Psychiatr. 2020;51:102127.
101. Pakpour AH, Griffiths MD. The fear of COVID-19 and its role in preventive behaviors. J Concurrent Disord. 2020;2(1):58–63.
102. Bhuiyan AI, Sakib N, Pakpour AH, Griffiths MD, Mamun MA. COVID-19-related suicides in Bangladesh due to lockdown and economic factors: case study evidence from media reports. Int J Ment Heal Addict. 2021;19:2110–5.
103. Dsouza DD, Quadros S, Hyderabadwala ZJ, Mamun MA. Aggregated COVID-19 suicide incidences in India: fear of COVID-19 infection is the prominent causative factor. Psychiatry Res. 2020;290:113145.

Post-Pandemic Burden of Non-communicable Disease

3

Vishal Chandra, Saurabh Mishra, and Rachna Dixit

3.1 Introduction

3.1.1 The Deadly Convergence: Non-communicable Diseases and COVID-19—A Global Health Crisis

The impact of COVID-19 on NCD worsens its condition and heightens overall health risks. During this pandemic, individuals have unmet medical facilities and desirable needs causing severe health and economic consequences that are prone to noncommunicable diseases. Enhanced healthcare policies specifically focused on integrating noncommunicable disease management within pandemic recovery strategies and ensuring access to care. The devastating condition of human health, especially for the low and middle-income ones, is grappling with three hurdles: undernutrition, communicable diseases, and a surge in noncommunicable diseases, during the COVID period, all of which result in significant health and economic repercussions, including increased healthcare costs and disability [1]. Irrefutably, the current pandemic has not just revealed the already vulnerable global health systems but widened what already exists as far as health inequities are concerned. These inequalities expose everyone to a flaming cascade of problems that affect populations in low- and middle-income countries the most. COVID-19 and NCDs converged to raise disastrous havoc on healthcare systems worldwide. Ongoing research indicates that by having pre-existing NCDs, one is at undergoing risk of being infected with severe COVID-19 complications and developing mortality. Studies already show that patients with heart problems, neurological symptoms, diabetes, or chronic respiratory illnesses, have been shown to have a threefold higher risk of developing serious illness [2]. The disruption of routine healthcare services has aggravated such risk, thereby delayed the work of treatment and diagnosis and damaged the overall disease management resiliencies which leads to mental stress [3]. A person already facing some health ailment, e.g., cardiovascular disease, high blood pressure, or diabetes, synergistically impacts health with COVID. Many socioeconomic factors affect public health whether physical, mental, and financial. Even if the vaccine eradicated COVID, its adverse consequences persist in the human body. People who already have noncommunicable diseases are more susceptible to COVID. The potential risk of infection is exacerbated by chronic diseases that compromise immune systems. Healthcare systems have become overwhelmed and unable to handle both issues: Fatal results from delayed diagnosis and treatment.

V. Chandra (✉) · S. Mishra · R. Dixit
Department of Life Sciences and Biotechnology,
Chhatrapati Shahu Ji Maharaj University,
Kanpur, Uttar Pradesh, India
e-mail: vishalchand@csjmu.ac.in

A. Raghav et al. (eds.), *COVID-19 Impact on Economy, Environment and Healthcare*,
https://doi.org/10.1007/978-981-95-1196-9_3

3.2 Major NCDS and Their Post-Pandemic Manifestations

3.2.1 Post-Pandemic Manifestations of Non-communicable Diseases

3.2.1.1 Cardiovascular Manifestations

The cardiovascular sequelae of post-COVID-19 syndrome present a complex constellation of symptoms that significantly impact patient outcomes. Myocarditis emerges as a notable complication, accompanied by various presentations of arrhythmia that necessitate careful clinical monitoring. The occurrence of thromboembolic disease represents a serious concern, potentially leading to long-term anticoagulation requirements. During the acute infection period, patients may experience acute myocardial injury, which can progress to heart failure in severe cases. The presence of persistent chest pain serves as a clinical indicator warranting thorough cardiovascular evaluation to prevent adverse outcomes. These manifestations collectively suggest a substantial burden on the cardiovascular system, potentially leading to chronic health implications that require sustained medical attention [4].

3.2.1.2 Respiratory System Manifestations

The respiratory system demonstrates a particularly vulnerable target for post-COVID-19 complications, characterized by a spectrum of functional and structural alterations. A significant finding includes impaired oxygen exchange mechanisms, manifesting as difficulty in oxygen removal from the pulmonary system. Patients frequently present with productive cough accompanied by sputum production, indicating ongoing inflammatory processes. The persistence of chest tightness and dyspnea significantly impacts quality of life, while abnormal lung imaging findings correlate with functional deficits. In severe cases, progression to respiratory failure necessitates ventilator support, highlighting the potential for long-term respiratory compromise. The combination of these symptoms, alongside abnormal lung function tests, suggests substantial respiratory system remodeling that may persist well beyond the acute phase of infection [5].

3.2.1.3 Neurological and Psychosomatic Manifestations

The neurological and psychosomatic manifestations of post-COVID-19 syndrome represent a significant burden on patient well-being, encompassing both cognitive and emotional domains. Mental health implications are particularly noteworthy, with patients experiencing a range of psychological disturbances including anxiety, depression, and post-traumatic stress disorder (PTSD). The development of obsessive-compulsive symptoms suggests potential long-term psychiatric sequelae. From a neurological perspective, patients exhibit various degrees of sensory and motor dysfunction, accompanied by persistent fatigue and muscle weakness. Cognitive changes manifest across multiple domains, potentially affecting daily functioning and occupational capabilities. The presence of dysphagia indicates potential involvement of cranial nerve function, while persistent headaches may suggest ongoing neurological inflammation. These diverse manifestations highlight the complex interplay between neurological and psychological systems in post-COVID-19 recovery [6] (Fig. 3.1).

The figure illustrates the spectrum of post-COVID-19 complications across three major physiological systems. Panel A depicts cardiovascular complications ($n = 7$), including chest pain, myocarditis, arrhythmia, thromboembolic disease, acute infection manifestations, acute myocardial injury, and heart failure. Panel B shows respiratory manifestations ($n = 10$), characterized by impaired oxygen exchange, sputum production, chest tightness, respiratory failure, persistent cough, dyspnea, abnormal lung imaging and function, chest pain, and ventilator dependence. Panel C presents psychosomatic and neurological manifestations ($n = 11$), encompassing anxiety, insomnia, fatigue, headache, depression, muscle weakness, PTSD, obsessive-compulsive symptoms, cognitive changes, sensory-motor dysfunction, and

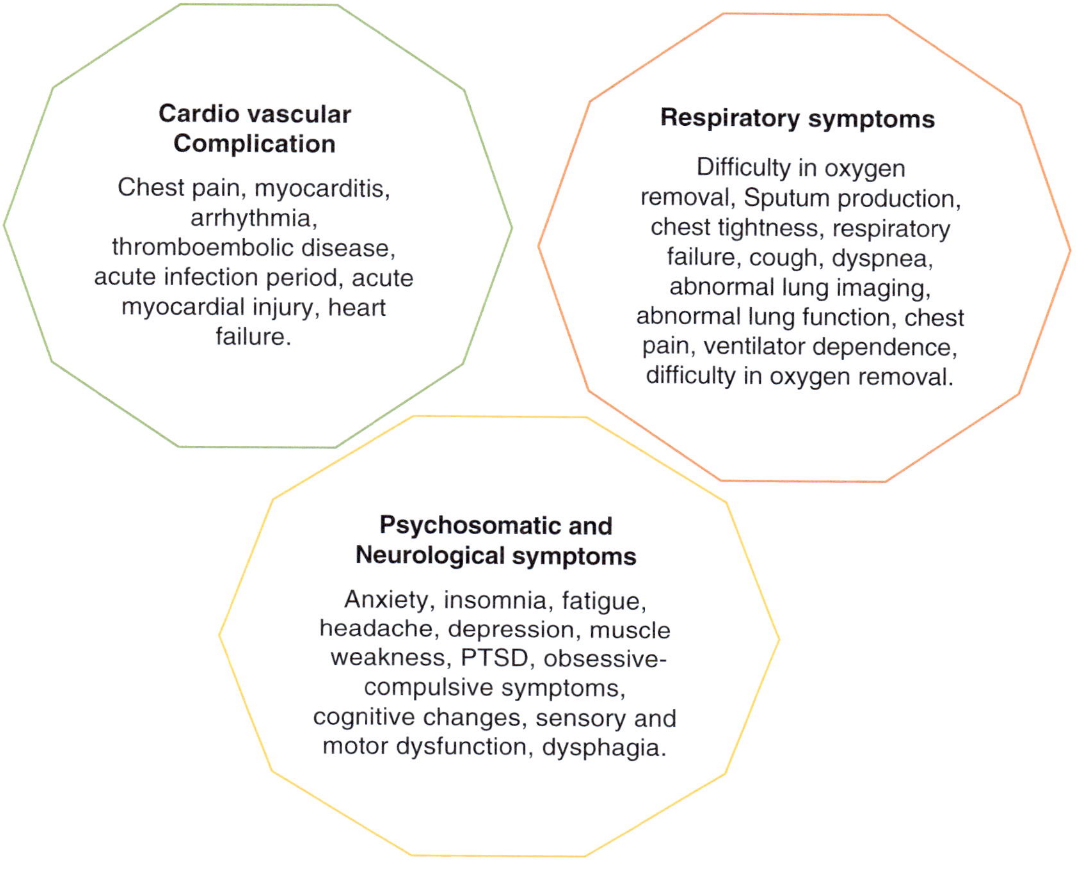

Fig. 3.1 Comprehensive overview of post-COVID-19 disease manifestations across major organ systems

dysphagia. This comprehensive symptom profile demonstrates the multisystemic nature of post-COVID-19 sequelae, highlighting the need for integrated clinical management approaches. Data sourced from clinical observations documenting the range of post-COVID complications across affected organ systems.

The diverse range of post-COVID-19 manifestations across cardiovascular, respiratory, and neurological systems necessitates a comprehensive, multidisciplinary approach to patient care. The chronic nature of these symptoms suggests the need for long-term follow-up protocols and specialized rehabilitation programs. Future research should focus on understanding the underlying mechanisms of these persistent symptoms to develop targeted therapeutic interventions. Additionally, the establishment of specialized post-COVID-19 clinics may facilitate better management of these complex manifestations and improve patient outcomes.

3.3 Healthcare System Disruptions and Adaptations

Due sudden outbreak of the COVID-19 pandemic created havoc upon the existing established healthcare system by disorganizing the systematic chain of essential healthcare services to the patients. Quality discrepancies of healthcare facilities between public and private healthcare providers are particularly evident in rural and urban areas, where access to resources, infrastructure, and medical personnel varies significantly. For the treatment of COVID-19, the basic need for oxygen supply and desirable medicine are not available in treatment areas. Many preliminary parameters are essential during the COVID period such as wearing masks, sanitizer usage, social distancing, avoiding going out unnecessarily, working from home, and only people of essential services (nurses, doctors, police, and home services) are exempted. During the first wave of COVID-19, there was a low infectivity since the lockdown was imposed and a person was practicing social distancing. However, in March 2021, the country faced the most lethal second wave as a consequence of individuals taking preventive measures casually. Increment in several cases resulted in a shortage of medicine, equipment supplies, and medical professionals. The healthcare system failed to meet the demands of the patients. With the emergence of the third wave of the new Omicron variant of SARS-CoV2 in January 2022, the cases have exponentially increased. The demand for oxygen beds, insufficient supplies (e.g., PPE), and hospital facilities was low and mostly remained unoccupied [2]. Public healthcare systems need to be restructured with better strategies, and initiatives have been taken to modernize and adapt healthcare systems (Fig. 3.2).

The image depicts a comprehensive set of global strategies employed during the COVID-19 pandemic to manage the public health crisis. At the core of these efforts was the goal of limiting the spread of the virus, which involved measures such as closing borders, restricting international traffic, and managing medical staff at entry points. Alongside these containment efforts, the strategies also focused on providing critical support to mitigate the negative economic impacts and scale up community-based approaches. Information dissemination played a key role, with public information campaigns and research promotion for vaccines and treatments. To protect frontline healthcare workers, the strategies included redistributing funds to purchase supplementary protective equipment and reorganizing workflows to avoid overburdening medical personnel. Importantly, the management component involved financial support for hospitals treating

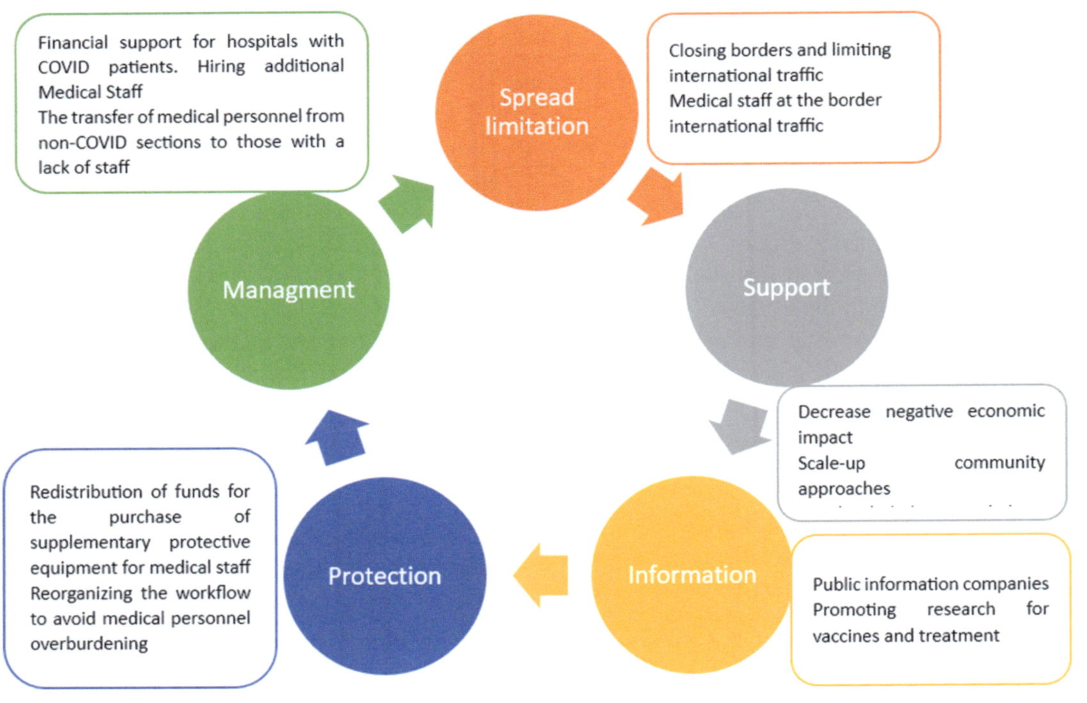

Fig. 3.2 Global strategies for pandemic management

COVID-19 patients, hiring additional staff, and transferring personnel from non-COVID sections to those facing shortages. This multifaceted, systems-level approach aimed to holistically address the complex public health, economic, and logistical challenges posed by the pandemic. By leveraging a diverse set of strategies, the global community sought to enhance pandemic preparedness and build more resilient healthcare infrastructures for the future. During post-COVID healthcare system needs to implement step-by-step Mondial strategies describing five measures that specifically focus on redesigning healthcare systems to better combat the future pandemic consequences: (1) restrict access to healthcare facilities to ensure patient and staff safety; (2) establish procedures and procedures to support, retain, and protect medical personnel; (3) transfer non-urgent hospital cases to outpatient clinics; (4) facilitate and promote communication between epidemiologists, virologists, point-of-care healthcare providers, and employees of healthcare facilities; (5) to ensure coordinated global action against COVID-19 and other emerging infectious diseases, best practice guidelines and laws should be developed. Healthcare provisions must be amended to better manage present and future pandemics, with effective planning and preparedness strategies executed to act as a shield against the rapid spread of pathogens and to ensure a swift, coordinated response. Reorganization or rebooting of the healthcare system can promote efficient healthcare with greater therapeutic availability, more effective ICU triage, expandable ICU capacity, adequate stocks of PPE, sufficient supply of consumables, telemedicine approaches, and advancement in treatment strategies [3–7].

3.3.1 COVID Vaccines and Their Side Effects

Global healthcare has experienced a significant transition as a result of the COVID-19 pandemic, which has shown serious vulnerabilities and emphasized the perseverance of medical innovations, especially vaccine programs.

The post-pandemic landscape reveals a complex interplay between healthcare system disruptions and the emergence of novel interventional approaches, with vaccines at the epicenter of this transformative period. These disruptions fundamentally altered disease management paradigms, especially for non-communicable diseases (NCDs) such as cardiovascular conditions, mental health challenges, and metabolic disorders. A disruption in routine healthcare services, carried out systematically, has had a cascading effect in delaying diagnostics and early interventions, and could adversely affect long-term health outcomes. Vaccinations have proven to be the most important intervention; they are preventive against COVID-19 as well as for protection from other health risks. Comprehensive studies across multiple vaccine platforms—including BNT162b2, ChAdOx1 nCoV-19, and mRNA-1273—demonstrated remarkable efficacy in mitigating post-acute symptoms and reducing systemic health complications [9]. The research revealed significant risk reductions: approximately 41% decrease in long-COVID severity, 34% reduction in mortality risks, and substantial improvements in multi-system health parameters. The vaccine's impact transcended traditional immunological protection, offering a multisystemic intervention strategy. By addressing symptoms across cardiovascular, neurological, metabolic, and psychological domains, vaccines provided a holistic approach to health management during unprecedented healthcare system disruptions. Clinical evidence consistently showed reduced fatigue, minimized muscular pain, improved cognitive functioning, and enhanced overall physiological resilience. Healthcare system adaptations became equally crucial in this transformative landscape. The pandemic accelerated digital health technologies, telemedicine platforms, and remote monitoring systems. These innovations emerged as critical solutions to address healthcare delivery challenges, particularly in managing non-communicable diseases that require continuous monitoring and personalized interventions. The intersection of vaccination strategies and healthcare system redesign revealed significant insights into population health management. Integrated approaches combining immunological protection, digital health technologies, and comprehensive patient monitoring demonstrated potential pathways for addressing complex health challenges in resource-constrained environments. Mental health emerged as a particularly critical domain, with vaccines and adaptive healthcare strategies showing promising results in mitigating psychological trauma, reducing stress-induced complications, and providing comprehensive support mechanisms. The pandemic highlighted the intricate connections between physiological and psychological health, necessitating more holistic healthcare approaches. Economic considerations further underscored the importance of these interventions. The potential productivity losses from unmanaged noncommunicable diseases, combined with the healthcare system's increased financial pressures, emphasized the need for cost-effective, innovative health management strategies [8] (Table 3.1).

Even if the vaccine eradicated COVID, its adverse consequences persist in the human body. People who already have noncommunicable diseases are more susceptible to COVID. Chronic illnesses that impair immune systems increase the possible risk of infection. Both problems have become too significant for healthcare systems to address. Postponed medical treatment and delayed diagnoses have led to deadly repercussions, especially as medical professionals continue to combat COVID-19, diverting resources and attention away from other critical health conditions. We must additionally develop extensive strategies to treat patients' symptoms and motivate them to get the medical assistance they need. This literature discusses whether the brain, heart, and lungs may be involved in post-viral syndromes and offers strategies for treating post-COVID-19 syndrome.

Post-COVID-19 respiratory manifestations comprise shortness of breath and coughing. Furthermore, arrhythmias, palpitations, hypotension, increased heart rate, venous thromboembolic diseases, myocarditis, and acute heart failure are usual cardiovascular events. Among neurological manifestations, headache, peripheral neuropathy symptoms, memory issues, lack of concentration, and sleep disorders are most commonly observed with varying frequencies. Finally, mental health issues affecting cognitive abilities and mood fluctuations, namely anxiety and depression, are frequently seen. Finally, long COVID is a complex syndrome with protracted heterogeneous symptoms, and patients who experience post-COVID-19 sequelae require personalized treatment as well as ongoing support [15].

According to data published by the World Health Organization, as of August 20, 2024, there were over 775 million confirmed cases of COVID-19 globally, which resulted in over 7 million fatalities [13]. Even when respiratory symptoms are absent, neurologic and cardiovascular symptoms have been identified as significant aspects of COVID-19, even though respiratory symptoms have been described as the most acute presentation of the disease [15–17].

Both adults and children may experience mild to severe neurological symptoms linked to COVID-19, which can happen during or after an acute SARS-CoV-2 infection. In the acute phase, the following neurological signs, symptoms, or syndromes have been reported: headache, ataxia, dizziness, fatigue, impaired taste or smell, muscle weakness, cognitive impairment, altered mental status/delirium, cerebrovascular complications, seizures, meningoencephalitis, encephalopathy, coma, and Guillain-Barré syndrome [15, 16]. The post-acute phase consequences, encompassing acute onset or chronic signs and symptoms (post-COVID-19 state), have been widely documented. These include, but aren't restricted to, cognitive impairment, headache, smell or taste issues, sleep disruptions, difficulty concentrating (often known as "brain fog"), drowsiness, and neuropsychiatric symptoms that fluctuate in intensity and duration [18]. Neurological conditions such as epilepsy, dementia, Parkinson's disease, and cerebrovascular illness are risk factors for the severity of COVID-19 and mortality during the acute infectious period [19]. During the initial phases of the pandemic, people with preexisting neurological conditions experienced difficulties getting care because of worries about

Table 3.1 Vaccine product impact on post-acute COVID-19 symptoms and health [14]

Vaccine information	Post-acute system reported	Impact of vaccine on symptoms associated with post-COVID	Author
Product: BNT162b2, ChAdOx1 nCoV-19, and mRNA 1273 Dose: Two doses Follow-up: At least 14 days after the first dose of vaccination and at least 7 days after the second dose of vaccination	Fever, persistent cough, loss of smell, fatigue, headache, sore throat, dizziness, chills or shivers, hoarse voice, brain fog, unusual muscle pains, eye soreness, diarrhea, shortness of breath, chest pain, nausea tinnitus, abdominal pain, earache	Product: BNT162b2, ChAdOx1 nCoV-19, and mRNA 1273 Dose: Two doses Follow-up: At least 14 days after the first dose of vaccination and at least 7 days after the second dose of vaccination	Antonelli et al. [9]
aOR (95%CI) Vaccinated: OR 0.65 (0.45–0.96) Unvaccinated: OR 0.55 (0.37–085)	Fatigue, joint pain, muscle, hair loss, headache breathlessness, cough, sleep disturbance	Product: Not reported Dose: Two doses Follow-up: Not reported	Senjam et al. [10]
Product: ChAdOx1 nCoV-19, BNT162b2, and mRNA 1273 Dose: Two doses Median follow-up Vaccinated: 96 days (IQR: 90–104) Unvaccinated: 98 days (IQR: 89–109)	ND	aOR (95%CI) Long-COVID of any severity: aOR 0.59 (0.50–0.69)	Ayoubkhani et al. [11]
Product: Ad26.COV2.S Dose: One Product: BNT162b2 Dose: Two Product: mRNA 1273 Dose: One Follow-up: Within 6 months	Cardiovascular, coagulation and hematologic, gastrointestinal, kidney mental health, metabolic musculoskeletal, neurologic disorders	BTI: Risk of death HR: 0.66 (0.58–0.74) burden of −10.99 (−13.45 to −8.22) Post-acute sequelae HR = 0.85 (0.82, 0.89) burden of −43.38 (−53.22 to −33.31) **negative values denote reduced burden in BTI relative to SARS-CoV-2 infection	Al-Aly et al. [12]
Product: BNT162b2, mRNA 1273 Ad26. COV2.S, unspecified subtype Dose: 1–2 Follow-up: Within 6 months	Abdominal symptoms, abnormal breathing, anxiety/depression, chest/throat pain, cognitive symptoms, fatigue, headache, myalgia, other pain	Fatigue (HR 0.89, 95% CI 0.81–0.97) Myalgia (HR 0.78, 95% CI 0.67–0.91) Pain (HR 0.90, 95% CI 0.81–0.99) Abnormal breathing (HR 0.89, 95% CI 0.81–0.98) Cognitive symptoms (HR 0.87, 95% CI 0.76–0.99) HR for other symptoms were not reported	Taquet et al. [13]

ND no data, *aOR* adjusted odds ratio, *SD* standard deviation, *OR* odds ratio, *HR* hazard ratio, *RR* risk ratio, *BTI* breakthrough infections infectious risks, disruptions in medicine supply chains, and resulting stock-outs. By the end of 2022, however, these issues had literally returned to normal.

One of the major adaptations in healthcare system is digital health solutions and telemedicine emerging as critical lifelines in maintaining continuous patient care and managing complex health challenges.

3.4 Digital Health Solutions and Telemedicine

The COVID-19 pandemic has catalyzed an unprecedented transformation in healthcare delivery, with digital health solutions and telemedicine it's more than temporary adaptations that act as life saviors for individuals and make it easy to consult with healthcare professionals. It signifies a fundamental reimagining of healthcare infrastructure, driven by necessity and technological innovation.

Technological infrastructure has rapidly evolved to address the multifaceted challenges of modern healthcare. Remote monitoring systems have emerged as pivotal tools, enabling healthcare professionals to track patient health continuously, detect early symptoms, and implement personalized intervention strategies. Artificial intelligence and advanced analytics have become cornerstone technologies in this digital health transformation. Machine learning algorithms now offer sophisticated capabilities in predictive risk assessment, generating personalized treatment recommendations, and enhancing diagnostic accuracy. These technologies transcend traditional medical approaches, providing dynamic, real-time insights into patient health trajectories and potential intervention strategies. The implementation of telemedicine involves a complex ecosystem of integrated digital consultation platforms, secure health data transmission mechanisms, and patient-centric design principles. Clinical evidence consistently demonstrates significant benefits, including reduced patient wait times, increased healthcare accessibility, and minimized infection transmission risks. Moreover, these platforms have proven particularly crucial in managing chronic diseases, enabling continuous patient engagement and real-time health parameter monitoring. Technological considerations extend beyond mere digital implementation. Robust cybersecurity protocols, high-speed internet connectivity, user-friendly interface designs, and comprehensive data privacy mechanisms have become essential components of effective digital health strategies [20]. The challenge lies not just in technological implementation but in creating holistic, accessible, and secure healthcare ecosystems. Challenges in digital health adoption are multifaceted, encompassing digital literacy variations, technology access disparities, data privacy concerns, and regulatory compliance complexities. Successful implementation requires a nuanced, phased approach that addresses these potential barriers through comprehensive training programs, healthcare professional digital skills development, and patient education initiatives. The regulatory landscape must simultaneously evolve to support these technological transformations. Clear telemedicine guidelines, robust data protection standards, and sophisticated quality assurance mechanisms are critical in establishing trust and ensuring effective digital healthcare delivery. This requires collaborative efforts between technological innovators, healthcare professionals, policymakers, and regulatory bodies [22]. Mental health management represents a particularly promising domain for digital health interventions. Telemedicine platforms offer unprecedented opportunities for continuous psychological support, breaking down traditional barriers of access and stigma. By providing flexible, confidential, and easily accessible mental health resources, these technologies can address the complex psychological impacts of pandemic-induced health disruptions. Global health collaboration has become increasingly important in this technological landscape. Sharing knowledge, resources, and innovative strategies can accelerate digital health adoption, particularly in resource-constrained environments. International cooperation can help develop adaptable technological frameworks customized to diverse healthcare contexts. Economic considerations further underscore the importance of digital health solutions. By reducing operational costs, minimizing unnecessary medical interventions, and enabling more efficient resource allocation, these technologies offer potential long-term economic benefits for healthcare systems globally. The convergence of vaccination strategies, digital health technologies, and adaptive healthcare models represents a transformative approach to population health management. These integrated strategies offer a comprehensive, proactive framework for addressing complex health challenges, moving beyond reactive treatment models to anticipatory, personalized healthcare interventions.

3.5 Policy Recommendations and Future Directions

India's strategy for dealing with non-communicable diseases (NCDs) changed dramatically during the COVID-19 pandemic, underscoring the crucial overlap between current health issues and the new global health emergency. People with pre-existing medical illnesses, especially those with noncommunicable diseases (NCDs), were more vulnerable to serious COVID-19 consequences as a result of the epidemic. India's non-communicable disease (NCD) policies were amended during the pandemic to deal with both immediate health issues and long-term strategic plans. As announced in 2022, the National Tele-Mental Health Programme gained unprecedented importance as a panacea through the pervasive mental health challenges triggered by the pandemic. Lockdowns, social isolation, and economic uncertainties created a mental health crisis needing innovative and accessible healthcare solutions. The Ayushman Bharat Programme was introduced in 2018 and became the need of the hour, as it proved significant during the pandemic because it provided robust healthcare infrastructure for emerging health needs at an accelerated pace. Its digital health infrastructure and package coverage became lifesavers for many poor communities suffering from both COVID-19 and other non-communicable diseases. The importance of fitness and nutrition has increased during the periods of the COVID-19 pandemic when individuals are restricted to confined places. The Fit India Movement and national fitness protocols during this time were major sources of resourceful movement to keep everyone healthy during this challenging phase. In the year of 2020, fitness and age-specific norms and procedures were formulated for children aged five to eighteen, adults aged eighteen to sixty-five years, and persons aged sixty-five and over. All these aspects thus ensured that persons received individualized recommendations for activity. The new regulations in terms of food safety and nutrition were enforced by the Food Safety and Standards Authority of India (FSSAI), particularly for school children to have balanced and safe food. These programs played a very important and affluent role for overall well-being and for benefiting the immune system during the difficult time of the pandemic. Furthermore, the update of the NNMS report on the National NCD Monitoring Survey provides data very critical at such a level for national NCD targets to enhance tracking and assessment of health outcomes against non-communicable diseases. Management of Healthcare systems underwent rapid transformation, with particular emphasis on managing most chronic conditions and neurological health conditions. The National Programme for Prevention and Control of Cancer, Diabetes, Cardiovascular Diseases, and Stroke (NPCDCS) demonstrated remarkable adaptability by implementing innovative strategies to ensure uninterrupted healthcare delivery for patients with chronic conditions while simultaneously mitigating COVID-19 transmission risks. Digital health systems and telemedicine have completely redefined health accessible—it makes the technological interventions essential for that. These infrastructures have closed the gap left by distance measures through further connection to remote consultations and continuous monitoring for medical purposes or personalized health management. In fact, patient-physician contacts in real-time were made easier through the use of cutting-edge telecommunication technologies, enabling reduced continuity to reach out even during unprecedented global outages. The pandemic's profound psychological implications necessitated a comprehensive revaluation of mental health strategies. The National Mental Health Policy, initially launched in 2017, transformed into a critical framework for addressing the exponential increase in psychological challenges triggered by prolonged uncertainty, social isolation, and economic instability. This policy framework provided a structured approach to understanding and mitigating mental health risks across diverse population segments. The National Multisectoral Action Plan for Prevention and Control of Common Non-Communicable Diseases (2017–2022) demonstrated strategic foresight by incorporating flexible response mechanisms. Complemented by the National Nutrition Strategy, this comprehensive approach emphasized holistic health management, recognizing the intricate interconnections between physical health, nutrition, and psychological well-being. Epidemiological evidence increasingly highlighted the complex interactions between chronic diseases, nutritional status, and psychological resilience. The multidisciplinary approach adopted by Indian healthcare policymakers reflected a sophisticated understanding of health as a multifaceted, interconnected system requiring integrated intervention strategies. The pandemic served as a catalyst for accelerating digital health transformations, emphasizing preventive healthcare, and developing robust, adaptable healthcare infrastructure. By prioritizing technological innovation, comprehensive policy frameworks, and patient-centric approaches, the healthcare system demonstrated remarkable resilience in navigating unprecedented global health challenges. This strategic response not only addressed immediate healthcare requirements but also laid a foundation for future healthcare models characterized by flexibility, technological integration, and holistic wellness perspectives. The lessons learned during this period will likely inform global healthcare strategies, emphasizing the importance of adaptive, comprehensive health management approaches. Climate change and environmental health programs, such as the National Programme on Climate Change and Human Health (NPCCHH) and the National Clean Air Programme (NCAP), took on new relevance as the pandemic highlighted the intricate connections between environmental factors and public health. The National NCD Monitoring Survey (NNMS) became even more critical in understanding the pandemic's impact on existing health conditions. The data collected during this period provided crucial insights into the changing health landscape and the interconnected nature of various health challenges.

Tobacco control efforts, including the Cigarettes and Other Tobacco Products Act (COTPA) and the National Tobacco Control Programme, gained additional significance as research emerged linking tobacco use to increased COVID-19 vulnerability [21]. The pandemic ultimately served as a catalyst, accelerating digital health innovations, emphasizing preventive healthcare, and demonstrating the need for a holistic, integrated approach to managing non-communicable diseases. India's response reflected a dynamic and adaptive strategy, transforming challenges into opportunities for healthcare system strengthening.

This comprehensive approach underscored the importance of resilient, flexible healthcare systems capable of addressing complex, interconnected health challenges in an increasingly unpredictable global environment (Fig. 3.3).

The figure depicts the current major national policies and programs addressing non-communicable diseases (NCDs) in India. The framework comprises four key initiatives: (A) The National Programme for Prevention of Cancer, Diabetes, Cardiovascular Diseases and Stroke (NPCDCS), which provides comprehensive NCD prevention and control; (B) The National Programme for Healthcare of Elderly (NPHCE), focusing on geriatric care and age-related NCDs [23]; (C) Ayushman Bharat's Health and Wellness Centers, offering primary healthcare services including NCD screening and management; and (D) The National Digital Health Mission, enabling digital health infrastructure for improved NCD care delivery. These programs collectively form India's current strategic response to the NCD burden.

The post-pandemic healthcare landscape stands at a critical inflection point, demanding a transformative vision beyond traditional medi-

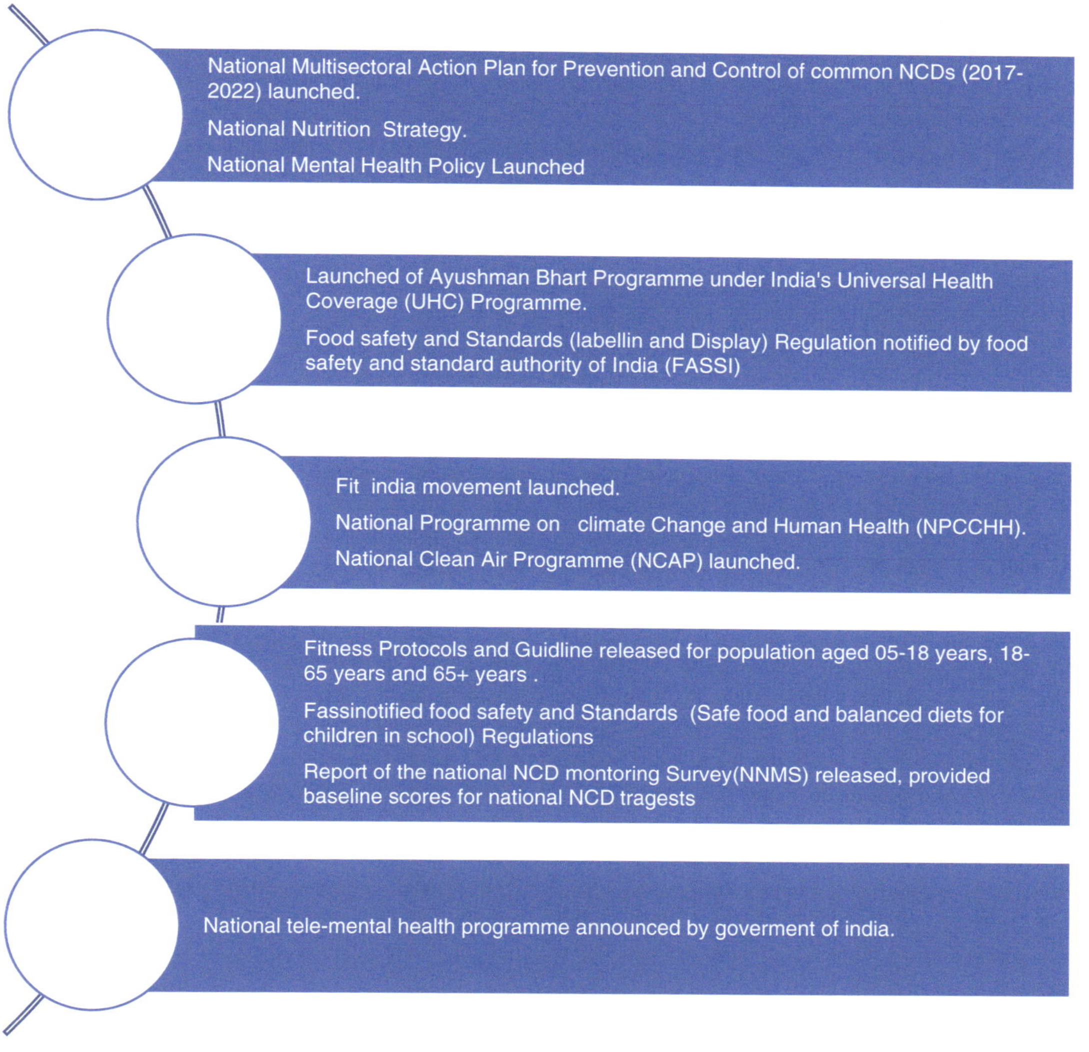

Fig. 3.3 Key national policies and programs for non-communicable diseases in India

cal paradigms. As we emerge from the unprecedented global health challenges of recent years, the future of healthcare requires a comprehensive, adaptive approach that integrates technological innovation, scientific understanding, and human-centered care strategies. Technological convergence emerges as a pivotal driver of future healthcare transformation. Artificial intelligence, advanced predictive analytics, and machine learning will transition from supplementary tools to fundamental healthcare infrastructure. These technologies promise to revolutionize medical practice by enabling unprecedented predictive capabilities, personalizing intervention strategies, and creating dynamic health management systems that can anticipate and respond to complex health challenges before they fully manifest. Vaccine research and development will undergo a fundamental paradigm shift, moving from reactive intervention to proactive health protection strategies. Future vaccine technologies will likely incorporate advanced genetic profiling, personalized immunological mapping, and multi-variant platform designs. The focus will expand beyond immediate viral prevention, exploring vaccines' potential to modulate long-term health trajectories, address complex systemic interactions, and provide comprehensive health protection. Global health collaboration will become increasingly critical in addressing complex, interconnected health challenges. International research networks, flexible health governance frameworks, and standardized global health protocols will emerge as essential mechanisms for sharing knowledge, resources, and innovative strategies. This collaborative approach will be particularly crucial in supporting healthcare systems in resource-constrained environments, ensuring more equitable global health outcomes. Healthcare professional education and workforce development require fundamental reimagination. Future healthcare professionals need to develop interdisciplinary competencies, advanced digital health skills, and the ability to navigate complex, technology-driven medical landscapes. Training programs must emphasize adaptability, systems thinking, and a holistic approach to health management that transcends traditional medical specialization boundaries. Ethical considerations and patient-centric design will become paramount in healthcare innovation. Future healthcare technologies must balance advanced capabilities with robust privacy protections, equitable access, and transparent governance. The development of health technologies will require a delicate balance between innovative potential and fundamental human rights, ensuring that technological advancements serve broader human well-being. Research priorities will expand to address the complex, multisystemic nature of health challenges revealed by the pandemic. Long-term studies will investigate the intricate interactions between physiological, psychological, and social health determinants. Researchers will explore the lasting impacts of global health disruptions, developing a more comprehensive understanding of human health resilience and adaptation mechanisms. Sustainability will emerge as a critical framework for future healthcare development. Healthcare systems must develop resource-efficient models, adaptive infrastructure, and comprehensive risk management strategies. This approach will require rethinking healthcare delivery from a holistic perspective, considering environmental, economic, and social factors that influence population health. The integration of digital health technologies will continue to accelerate, creating more personalized, accessible, and efficient healthcare delivery models. Telemedicine, remote monitoring, and AI-driven diagnostic tools will become standard components of healthcare ecosystems, breaking down traditional barriers of access and creating more responsive, patient-centered care approaches. Mental health will receive unprecedented attention, recognized as a critical component of comprehensive health management. Future healthcare strategies will develop more sophisticated, integrated approaches to psychological well-being, leveraging technology to provide continuous, personalized mental health support and addressing the complex psychological impacts of global health challenges [24]. The post-pandemic era is not an endpoint, but a beginning—an extraordinary opportunity to create more resilient, adaptive, and compassionate approaches to global health management. Our collective challenge lies in translating these visionary concepts into tangible, implementable strategies that can improve human health and well-being on a global scale.

References

1. Haileamlak A. The impact of COVID-19 on non-communicable diseases. Ethiop J Health Sci. 2022;32(1):1.
2. Davis HE, McCorkell L, Vogel JM, Topol EJ. Long COVID: major findings, mechanisms and recommendations. Nat Rev Microbiol. 2023;21(3):133–46.
3. Menting J, van Schelven F, Aussems C, Heijmans M, Boeije H. Routine healthcare disruptions: a longitudinal study on changes in self-management behavior during the COVID-19 pandemic. BMC Health Serv Res. 2023;23(1):196. https://doi.org/10.1186/s12913-023-09119-x.
4. Tanni SE, Tonon CR, Gatto M, Mota GA, Okoshi MP. Post-COVID-19 syndrome: cardiovascular manifestations. Int J Cardiol. 2022;369:80.
5. Al-Jahdhami I, Al-Naamani K, Al-Mawali A, Bennji SM. Respiratory complications after COVID-19. Oman Med J. 2022;37(1):e343.
6. Jung YH, Ha EH, Park J, Choe KW, Lee WJ, Jo DH. Neurological and psychiatric manifestations

of post-COVID-19 conditions. J Korean Med Sci. 2023;38(11):e83.
7. Andrews N, Stowe J, Kirsebom F, Toffa S, Rickeard T, Gallagher E, et al. Covid-19 vaccine effectiveness against the Omicron (B. 1.1.529) variant. N Engl J Med. 2022;386(16):1532–46.
8. Antonelli M, Pujol JC, Spector TD, Ourselin S, Steves CJ. Risk of long COVID associated with delta versus omicron variants of SARS-CoV-2. Lancet. 2022;399(10343):2263–4.
9. Simon MA, Luginbuhl RD, Parker R. Reduced incidence of long-COVID symptoms related to administration of COVID-19 vaccines both before COVID-19 diagnosis and up to 12 weeks after. MedRxiv. 2021;2021-11
10. Antonelli M, Penfold RS, Merino J, Sudre CH, Molteni E, Berry S, et al. Risk factors and disease profile of post-vaccination SARS-CoV-2 infection in UK users of the COVID Symptom Study app: a prospective, community-based, nested, case-control study. Lancet Infect Dis. 2022;22(1):43–55.
11. Senjam SS, Balhara YPS, Kumar P, Nichal N, Manna S, Madan K, et al. Assessment of post COVID-19 health problems and its determinants in North India: a descriptive cross section study. medRxiv. 2021;2021-10
12. Ayoubkhani D, Bosworth ML, King S, Pouwels KB, Glickman M, Nafilyan V. Risk of long covid in people infected with SARS-CoV-2 after two doses of a COVID-19 vaccine: community-based, matched cohort study. medRxiv preprint. 2022;10(2022.02):23–22271388.
13. Visco V, Vitale C, Rispoli A, Izzo C, Virtuoso N, Ferruzzi GJ, et al. Post-COVID-19 syndrome: involvement and interactions between respiratory, cardiovascular and nervous systems. J Clin Med. 2022;11(3):524.
14. Notarte KI, Catahay JA, Velasco JV, Pastrana A, Ver AT, Pangilinan FC, et al. Impact of COVID-19 vaccination on the risk of developing long-COVID and on existing long-COVID symptoms: a systematic review. EClinicalMedicine. 2022;53:101624. https://doi.org/10.1016/2Fj.eclinm.
15. Fridman S, Bres Bullrich M, Jimenez-Ruiz A, Costantini P, Shah P, Just C, et al. Stroke risk, phenotypes, and death in COVID-19: systematic review and newly reported cases. Neurology. 2020;95(24):e3373–85.
16. Favas TT, Dev P, Chaurasia RN, Chakravarty K, Mishra R, Joshi D, et al. Neurological manifestations of COVID-19: a systematic review and meta-analysis of proportions. Neurol Sci. 2020;41:3437–70.
17. Abdullahi A, Candan SA, Abba MA, Bello AH, Alshehri MA, Afamefuna Victor E, et al. Neurological and musculoskeletal features of COVID-19: a systematic review and meta-analysis. Front Neurol. 2020;11:687.
18. Panda PK, Sharawat IK, Panda P, Natarajan V, Bhakat R, Dawman L. Neurological complications of SARS-CoV-2 infection in children: a systematic review and meta-analysis. J Trop Pediatr. 2021;67(3):fmaa070.
19. Post-COVID-19-condition [Internet]; 2023 [cited 05.09.2023]. Available from: https://www.who.int/teams/healthcare-readiness/post-covid-19-condition
20. Boruah AP, Thakur KT, Gadani SP, Kothari KU, Chomba M, Guekht A, et al. Pre-existing neurological conditions and COVID-19 co-infection: data from systematic reviews, meta-analyses, and scoping reviews. J Neurol Sci. 2023;455:120858.
21. Li JPO, Liu H, Ting DS, Jeon S, Chan RP, Kim JE, et al. Digital technology, tele-medicine and artificial intelligence in ophthalmology: a global perspective. Prog Retin Eye Res. 2021;82:100900.
22. Kapoor M, Nidhi Kaur K, Saeed S, Shannawaz M, Chandra A. Impact of COVID-19 on healthcare system in India: a systematic review. J Public Health Res. 2023;12(3):22799036231186349.
23. Vaishnav LM, Joshi SH, Joshi AU, Mehendale AM. The national programme for health care of the elderly: a review of its achievements and challenges in India. Ann Geriatr Med Res [Internet]. 2022;26(3):183–95. Available from: https://doi.org/10.4235/agmr.22.0062.
24. Filip R, Gheorghita Puscaselu R, Anchidin-Norocel L, Dimian M, Savage WK. Global challenges to public health care systems during the COVID-19 pandemic: a review of pandemic measures and problems. J Pers Med. 2022;12(8):1295.

Post-COVID-19 Burden of Communicable Diseases

4

Rachna Dixit, Saurabh Mishra, Vishal Chandra,
Renu Tomar, Poonam Khanna,
Qazi Noorul Mateen, Romana Rashid,
and Shahid Banday

4.1 Introduction

The severe acute respiratory syndrome coronavirus 2 (SARS-CoV-2) pandemic, which began in February 2021, has cost healthcare systems and society around the world a great deal of money and caused over 110 million infected cases and about 2.5 million fatalities [1]. Although the initial focus of public health responses was on reducing the acute burden of COVID-19, there is increasing proof that SARS-CoV-2 infection can also have longer-term effects on physical and mental health, which are becoming a greater concern for healthcare systems [2–4]. The terms "post-COVID-19 syndrome" and "Long Covid" are currently used to describe such effects that persist for more than 3 months following infection [4]. The WHO and UNICEF child survival strategy, the WHO Integrated Management of Childhood Illness, and the integrated Global Action Plan for Prevention and Control of Pneumonia and Diarrhoea (GAPPD) are important programs that have contributed to these improvements. The Millennium Development Goals also focused on HIV, malaria, and other communicable diseases at the community level (goal 6), as well as on vaccine-preventable diseases like diarrhoea and pneumonia in early children (goal 4). This theme was carried over into the Sustainable Development Goals (SDGs), where one of the goals is to eliminate avoidable deaths among children under five (target 3.2), combatting hepatitis, water-borne infections, and other infectious diseases; and putting an end to the epidemics of AIDS, TB, malaria, and neglected tropical diseases (goal 3.3). International commitments, global policy, and national action against these targets have been fuelled by global funding initiatives such as the Global Fund to Fight AIDS, Tuberculosis and Malaria (HIV, TB, and malaria), the US Presidents' Emergency Plan for AIDS Relief, the

R. Dixit · S. Mishra · V. Chandra
Department of Life Sciences and Biotechnology, Chhatrapati Shahu Ji Maharaj University, Kanpur, Uttar Pradesh, India
e-mail: vishalchand@csjmu.ac.in

R. Tomar
School of Public Health, Poornima University, Jaipur, Rajasthan, India

P. Khanna
School of Public Health, Post Graduate Institute of Medical Education and Research, Chandigarh, India

Q. N. Mateen
Liverpool John Moores University, Liverpool, UK

R. Rashid
Department of Medicine, University of Massachusetts Chan Medical School, Worcester, MA, USA
e-mail: romana.rashid1@umassmed.edu

S. Banday (✉)
Department of Molecular, Cell and Cancer Biology, University of Massachusetts Chan Medical School, Worcester, MA, USA
e-mail: shahidkhursheed.banday@umassmed.edu

A. Raghav et al. (eds.), *COVID-19 Impact on Economy, Environment and Healthcare*,
https://doi.org/10.1007/978-981-95-1196-9_4

Global Alliance for Vaccines and Immunization, and the Grand Challenges program through the Bill & Melinda Gates Foundation [2, 4]. Lock down, quarantine, and other preventive measures have made it difficult for the general population to get a complete diagnosis of diseases other than COVID-19. A person suffering from any communicable disease is more prone towards the SARS-CoV-2 infection. Compared to 2019, 29% fewer persons received a TB test, according to Global Fund surveys conducted in 13 of the world's most TB-affected nations. Even worse, multidrug-resistant TB, one of the most terrifying types of antibiotic resistance, was tested for 45% fewer people in those same nations [3]. Because of overlapping symptoms and diagnostic challenges, diagnosing co-infection can be difficult. Like tuberculosis, COVID-19 can cause respiratory symptoms as coughing and dyspnoea. There is molecular testing for SARS-CoV-2 and sputum or molecular tests for TB, but the diagnostic procedures for COVID-19 and TB differ [5]. Due to the COVID-19 pandemic, people are not taking any initiative to diagnose their minor ailments and diagnoses are not fully done which result in severe health problems. Patients with TB often remain undiagnosed as the number of sputum sample submissions declined, with many avoiding healthcare facilities owing to early COVID-19 restrictions and fear of infection [6, 60]. Figure 4.1 shows the correlation between poverty and population density, healthcare expenditure, communicable diseases, and global depression.

The interconnected nature of these factors, as illustrated in the model, underscores the complex challenges faced during the COVID-19 pandemic. Addressing this crisis requires a com-

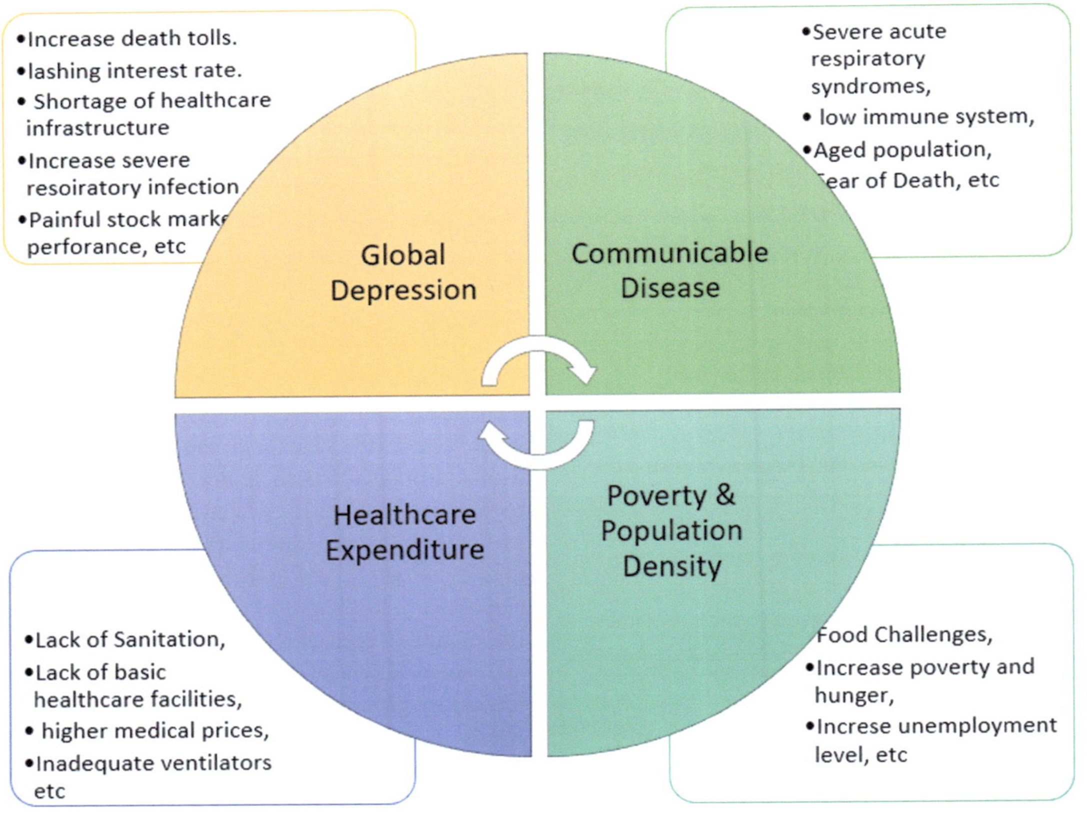

Fig. 4.1 The correlation between poverty and population density, healthcare expenditure, communicable diseases, and global depression

prehensive, multifaceted approach that considers the dynamic relationships between global depression, communicable disease characteristics, healthcare system capacity, and socioeconomic conditions. When the pandemic erupted, NPIs were implemented worldwide to stop the virus's infection [7–9]. Social distancing strategies, e.g., lockdowns, quarantine, travel restrictions, and personal hygiene measures (gloves, face masks, and sterilized infected surfaces) were among them [7, 8, 10]. Vaccination programme implementations have since resulted in the relaxation of non-pharmacological interventions; however, it is still unknown how the pandemic will ultimately affect vaccine-preventable diseases worldwide [9]. COVID-19 measures may not have been the only factors contributing to the substantial suppression of vaccine-preventable diseases (VPDs) like measles during the pandemic; pre-existing vaccination efforts and fewer medical visits might have played an important role. It explores the wide-ranging impacts of the pandemic on healthcare systems, epidemiology, and public health strategies, while identifying key lessons and recommendations for future preparedness. By examining these interconnected factors, the chapter aims to shed light on how global health systems can adapt to the challenges of both pre- and post-COVID pandemics.

4.1.1 Global Impact of the COVID-19 Pandemic on Infectious Diseases

The COVID-19 pandemic has precipitated a profound reconfiguration of infectious disease epidemiology. Global surveillance data indicates substantial shifts in transmission patterns, healthcare utilization, and disease prevalence across multiple infectious disease categories [11, 12]. Preliminary analyses suggest that pandemic-induced disruptions have created a dynamic and potentially volatile communicable disease environment. The Global Fund had a remarkable year in 2021. It provided US$3.2 billion to nations through its COVID-19 Response Mechanism (C19RM), in addition to contributing US$4.2 billion annually to combat HIV, TB, malaria, and improve health systems. It improves pandemic readiness and makes personal protective equipment (PPE), COVID-19 tests, and more accessible [13]. Global infectious disease epidemiology was substantially impacted in numerous ways by the COVID-19 pandemic. The manifestation of the epidemic was an attenuation of the incidence rates of several infectious diseases, particularly those with a respiration droplet- and contact-transmitted pathophysiology, owing to the combined effect of several NPIs (non-pharmaceutical interventions) like mask use, social distancing, closure, and travel bans.

Table 4.1 comprehensively compares various communicable diseases in terms of their main mode of transmission, availability of vaccination, the specific disease or infection, and how they compare to the pandemic in terms of severity. It covers a vast range of communicable diseases, including those transmitted through faecal-oral, sexually transmitted, blood-borne, vector-borne, and airborne routes. It also includes information on diseases transmitted through skin wounds, contaminated food or water, and direct contact. Regarding the COVID-19 pandemic, this table highlights how certain communicable diseases, such as Hepatitis A, Hepatitis B, and Viral meningitis, are considered lower in comparison, while others, like Legionellosis, are higher in severity. The table also reveals that some diseases, such as Tuberculosis and Pertussis, are comparable to COVID-19 in terms of their characteristics and impact. This knowledge can help one comprehend the relative difficulties presented by various infectious diseases, especially for post COVID-19 pandemic and its major impact on worldwide health. The most serious direct effects of COVID-19 were on the monitoring and reporting of many infectious diseases worldwide. Such global decline in disease notifications during this period might have been due to many other factors including underreporting, change in healthcare-seeking behaviour, and reduced access to healthcare. Many people due to COVID-19 fear or lockdown are not ready to diagnose minor ailments, which leads to suffering from major health

Table 4.1 Main mode of transmission, vaccination availability, disease/infection, and compared to pre-COVID-19

Main mode of transmission	Vaccination availability	Disease/infection	Compared to pre-COVID-19
Faecal-oral transmission	Yes	Hepatitis A	Lower
Sexually transmitted and via body fluids	Yes	Hepatitis B	Comparable
Blood-borne	No	Hepatitis C	Lower
Faecal-oral transmission	No	Hepatitis E	Comparable
Vector-borne	No	Visceral leishmaniasis	Comparable
Transmission through skin wound	Yes	Tetanus	Comparable
Airborne transmission	No	Viral meningitis	Comparable
Airborne transmission	Yes	Tuberculosis	Lower
Sexually transmitted and via body fluids	No	Syphilis	Comparable
Direct contact	No	Pediculosis	Comparable
Direct contact	No	Scabies	Comparable
Airborne transmission	No	Scarlet fever	Lower
Contaminated food or water	No	Food poisoning	Comparable
Faecal-oral transmission	No	Brucellosis	Lower
Sexually transmitted by via body fluids	No	HIV	Lower
Airborne transmission	Yes	Invasive bacterial disease (IBDs)	Lower
Inhalation of contaminated aerosol	No	Legionellosis	Higher
Contaminated food or water	No	Non-typhoid salmonellosis	Comparable
Airborne	Yes	Rubella	Comparable
Vector borne	No	Malaria	Lower
Airborne transmission	Yes	Pertussis	Comparable
Airborne transmission	Yes	Mumps	Lower
Airborne transmission	Yes	Chickenpox	Lower
Transplacental	No	Congenital syphilis	Lower

problems such as chickenpox and scarlet fever which are not diagnosed properly. Before COVID-19, syphilis, gonorrhoea, and chlamydia rates had been rapidly increasing in the United States. In 2019, the Centres for Disease Control and Prevention (CDC) stated that the rates of chlamydia, gonorrhoea, and syphilis increased by 3%, 5%, and 14%, respectively, with a 40% increase in congenital syphilis in infants [14, 15]. Legionella pneumophila co-infections are found in patients with severe acute respiratory syndrome coronavirus 2 in most cases [16]. The COVID-19 pandemic has caused serious flaws in illness detection and control procedures, highlighting the necessity of stronger and more durable worldwide monitoring systems to prepare for future pandemics and stop the reappearance of other infectious diseases.

4.2 Revitalizing Health Systems Amid COVID-19 Disruptions

4.2.1 Epidemiological Transformations

Preliminary measures like masks, sanitisers, and social distancing are considered during the COVID-19 pandemic. During 2020–2021, significant declines in influenza infection rates were observed; in certain areas, seasonal influenza prevalence decreased by up to 98% [17, 18]. Digital health software was designed during this pandemic to assist in tracking the outbreak of COVID, diagnose suspected patients, and speed up the finding of vaccines and disinfecting the surroundings. In April 2020, the COVID-19 vaccinations Global Access (COVAX) program was established in response to the early recognition of the need for fair access to COVID-19 vaccinations. COVAX is a collaborative effort led by the World Health Organization (WHO), the Coalition for Epidemic Preparedness Innovations (CEPI), and the Global Vaccine Alliance (Gavi). The goal of the COVAX initiative was to coordinate global resources to support the development and production of the various COVID-19 vaccines while ensuring that Lower-Middle-Income Countries (LMICs) and Low-Income Countries (LICs) had fair access to COVID-19 tests, treatments, and vaccines by acting as a buyers and distributor club through donations or self-financing schemes [62]. COVAX has established the PAHO Revolving Fund as the mechanism for acquiring the vaccines on behalf of countries in the region [63].

In response to this pandemic, the Access to COVID-19 Tools (ACT) Accelerator was introduced in April, using COVAX as one of its three pillars, bringing together governments, international health organizations, manufacturers, scientists, the corporate sector, civil society, and charity to provide COVID-19 diagnoses, treatments, and vaccinations in a way that is both creative and equitable. The latter is the main emphasis of the COVAX pillar. Because it is the sole attempt to guarantee that everyone in the globe, regardless of wealth, will have access to COVID-19 vaccines once they become available, it is the only truly global solution to this epidemic. While next-generation sequencing (NGS) is a technology that is utilized in many labs worldwide to examine the genetic composition of all living things, its application in the diagnosis of infectious diseases is currently rare. A genetic knowledge can be obtained from NGS, which determines the pathogen's genome sequence, then from standard testing methods. This information can be used to develop vaccines and treatments, track modifications to the virus as it spreads throughout the population, and gain a deeper understanding of patterns of transmission over time and space [19–22]. Global partnerships like the COVID-19 and the Global Initiative on Sharing All Influenza Data (GISAID) and Genomics UK Consortium (COG-UK) made it easier to share genomic data rapidly, which helped identify important mutations in variants like Alpha, Delta, and Omicron and monitor their global spread [23–25].

4.2.2 Management of Adaptive Strategies and Challenges in Healthcare Infrastructure

When SARS-CoV-2 first appeared in December 2019, the globe encountered a serious new public health threat that spread so drastically within a matter of weeks that most countries across the world were affected, and the major drawback is the medical care infrastructure which is not well equipped to manage the pandemic situation. Global healthcare system faced unprecedented resource allocation challenges which include overoccupied hospital capacity and medical equipment shortages and PPE (personal protective kit) shortage. Telemedicine is limited, communications systems are outdated, and diagnostic capacities are not up to the mark. During the pandemic, healthcare facilities experienced major financial challenges due to a significant rise in COVID-19-related expenditures and a decline in revenue from ordinary medical services. Technological shortcomings make the situation worse which hinders the ability of many institutions to manage the crisis more effectively. To act on this drastic situation, healthcare systems created novel adaptable strategies in response to these difficulties, which revolutionized the management of infectious diseases. AI-driven disease surveillance, diagnostic technology, and telemedicine platforms have advanced because of the pandemic's attention to the need for innovation in healthcare. Flexible healthcare delivery models, decentralized services, and enhanced interdepartmental coordination became essential. New pandemic preparedness protocols were developed, including infection control standards, strategic stockpiling of medical resources, and cross-institutional collaboration frameworks. In addition to enhancing data exchange and utilizing machine learning for predictive analytics, the swift digital transition included electronic health records. Experts like Haldane et al. [27] and Remuzzi and Remuzzi [28] have conducted research that highlights the significance of the healthcare system's adaptability. Main priorities are to work on healthcare infrastructure weaknesses and provide the vaccines and treatments to all people equally [26–28].

4.3 Epidemiological Analysis of Communicable Diseases

The significant transformation has been undergone in the field of epidemiology during post-COVID era. Researchers are using methodical techniques to assess the frequency of infectious diseases and monitor minute changes in the mechanisms of transmission across various groups. Consequently, the worldwide burden of illness and the mortality rates for COVID-19 and other health conditions are rising [29, 30]. In countries with a high burden of tuberculosis, mathematical models predicted increased mortality due to significant disruptions in antiretroviral treatment services. Projections also indicated that a 75% interference with the control of malaria efforts drastically reduces treatment and testing rates [31]. The weaknesses in paediatric and critical care services across Africa have resulted in high in-hospital mortality rates among children with COVID-19, reaching 8.3% compared to approximately 1% in high-income countries [32]. Unless the pandemic is effectively contained, it is expected to continue hindering HIV, TB, and malaria control efforts [30, 31]. The Global Fund to Fight AIDS, Tuberculosis, and Malaria (GFATM) has warned that COVID-19 could reverse decades of progress in combating these diseases. Additional research is essential to better understand how COVID-19 prevention measures are influencing the transmission of infectious diseases in Africa and other regions. The economic downturn caused by the COVID-19 pandemic has significantly impacted households by reducing incomes and increasing unemployment rates [33]. This emphasizes the urgent need for countries to strengthen social protection systems to ensure the well-being of their populations. Unfortunately, at the national level, many African countries were forced to redirect resources from essential services and take on additional debt to combat the pandemic. Combined with reduced economic activity and shrinking fiscal space, governments now face limited options to meet the growing demands for health and social support systems [30, 33]. To address these challenges, global solidarity is essential for sustainable and equitable investments in combating TB, HIV, and malaria, while also preparing for ongoing and future health threats. Policymakers must prioritize resource allocation and investments to strengthen health systems. Collaborative, cross-country studies are also needed to produce detailed data on disease surveillance, comorbidities, outcomes, and costs to better inform pandemic preparedness and response strategies. Funding organizations, both national and international, should focus on advancing epidemiological research, molecular diagnostics, and surveillance systems to enhance early detection and response capabilities. Frontline healthcare workers should leverage point-of-care diagnostic tools to improve clinical decision-making and deliver appropriate COVID-19 care. The unequal distribution of COVID-19 vaccines has highlighted the need for urgent investments in vaccine manufacturing hubs in Africa. Expanding these facilities will not only address current inequities but also ensure the region is better prepared for future pandemics [32, 34, 35]. While the pandemic appeared to stabilize in some regions, the disruption of routine health services has reversed progress in key areas. Moving forward, nations must embrace innovation and foster stronger collaboration to rebuild healthcare systems and resume progress towards achieving health goals. Innovation, collaboration, human rights, and scientific advancements are more critical than ever in addressing health challenges. To ensure continuous access to essential services, increased domestic and international investments are necessary [36]. Technological innovations are particularly valuable in improving service delivery, including enhancing data systems, ensuring the supply of medical essentials, enabling e-learning, strengthening disease surveillance, and supporting effective supervision [37]. A notable example is the RTS, S malaria vaccine, which has proven to save one life for every 200 children vaccinated, reduce malaria cases by 40%, and significantly lower severe malaria-related deaths. This vaccine can be delivered through established childhood immunization platforms, which already cover over 80% of children [36]. To maintain access to high-quality HIV and TB services during the COVID-19 pandemic, practical strategies include implementing community-based service delivery models, minimizing the frequency of healthcare visits and medication pickups, expanding mental health support, and prioritizing a stronger, more resilient health system. Placing people-centred care at the forefront of health services will help address the evolving needs of communities and improve health outcomes. During the COVID-19 pandemic, innovative strategies were employed to ensure continuity of HIV care. In countries like Nigeria, Indonesia, Laos, and Nepal, antiretroviral therapy (ART) was delivered to people living with HIV through community networks or private couriers, preventing disruptions in treatment [39]. These examples highlight how local solutions can effectively sustain access to essential medicines in challenging times. To better handle future pandemics, health systems must prioritize equity, preparedness, and resilience, ensuring the continuity of critical services to save lives [40]. The One Health approach serves as a vital framework for addressing zoonotic threats, integrating human, animal, and environmental health sectors to strengthen response strategies [41, 42]. At the same time, it is crucial not to overlook other high-burden infectious diseases in Africa, which continue to strain health systems alongside the challenges posed by COVID-19 [43].

4.4 Impact on Specific Communicable Diseases: A Comprehensive Narrative Analysis

The COVID-19 pandemic has reshaped our understanding of communicable diseases, exposed vulnerabilities and driven new research opportunities. The post-pandemic era has transformed infectious disease dynamics across respiratory, vector-borne, sexually transmitted, and vaccine-preventable diseases. Respiratory infections like influenza and RSV show complex transmission patterns influenced by pandemic measures and improved hygiene practices. Vector-borne diseases, such as malaria and dengue, are spreading to new regions due to climate change and ecological shifts. Sexually transmitted infections, including HIV and syphilis, faced setbacks from reduced testing, while telemedicine emerged as a critical tool for management. Vaccine-preventable diseases saw declining immunization rates, exposing gaps in healthcare equity and trust. The interconnected nature of these challenges highlights the need for interdisciplinary approaches, precision epidemiology, and technological innovations. The pandemic has driven fundamental changes in how infectious diseases are understood, managed, and prevented. Pandemic response measures, including lockdowns, significantly disrupted tuberculosis (TB) diagnosis and treatment services. In 2020, global TB detection rates dropped by 29% as fewer sputum samples were collected due to reduced healthcare access and the prioritization of COVID-19 over other conditions [3, 6]. High-burden countries experienced severe declines in TB notifications, with reductions of 41% in South Africa and 25% in India [44]. In Africa, which includes 17 countries with the highest TB burden, diagnostic and screening programs were disrupted by workforce shortages, healthcare facilities limited access, and issues with data documentation. According to the WHO, only 5.8 million of the 10 million TB cases that occurred in 2020 were identified, leaving 4.2 million cases unattended, which contributes to continued transmission within the community [45]. TB-related deaths rose from 1.2 million (2019) to 1.3 million (2020) among the HIV-negative group and from 209,000 to 214,000 among the HIV-positive group. Without urgent intervention, undetected TB cases are likely to fuel further transmission and mortality [45]. The primary cause of tuberculosis includes undernutrition, poverty, and stress, with untreated cases potentially fatal for more than half of those affected

[46]. TB and COVID-19 co-infections have been shown to triple mortality rates compared to TB alone [47]. Disruptions to healthcare services, coupled with COVID-19 stigma, have disproportionately impacted low-income communities, leading to higher TB incidence and mortality [48]. Despite these challenges, TB and COVID-19 co-infections remain limited in many African countries. Main cross-continental studies are urgently needed basically to assess undiagnosed TB cases, trends in drug-resistant TB (MDR-/XDR-TB), and the co-infection's impact on treatment results [49]. Additionally, pharmaceutical supply chain disruptions and national Tuberculosis programs need immediate intervention to prevent further setbacks. According to the Global Fund, compared to 2019, there was a 41% drop in HIV testing and a 37% drop in referrals for diagnosis and treatment in 2020 [3].

Modelling by Jewell et al. [50] showed that disruptions in antiretroviral drug (ART) supply could negatively impact HIV/AIDS trends, with a 50% reduction in condom availability potentially increasing new infections by 1.19 times and service disruptions leading to a 1.06 (fold) rise in HIV-related deaths. Simulation models suggested that maintaining HIV services could prevent 19-146 HIV deaths per 10,000 clients, with a minimal increase in COVID-19 deaths [50, 51]. In South Africa, during the 2020 national lockdown specifically in KwaZulu-Natal, where 1.7 million people live with HIV disease, ART provision was largely maintained, but HIV testing and ART initiations were significantly impacted [52]. Lessons learned include prioritising essential services, integrating health services, supplying prescriptions for chronic drugs that last several months, and ensuring continuous patient education. National HIV programs, WHO, and local organizations must make regulated changes to maintain essential services, expand testing, and initiate treatments, especially during COVID-19 surges in low-income and middle-income countries. According to WHO estimates, there will be 241 million cases of malaria and 627,000 fatalities worldwide in 2020, which is 14 million more cases and 69,000 more deaths than in 2019. Of these, 47,000 (deaths) were attributed to disruptions in malaria diagnosis prevention, and treatment due to COVID-19 [38]. In Africa, deaths rose by 12% between 2019 and 2020 [38], and models predict that the situation may worsen in 2021 [31]. Despite progress with Rapid Diagnostic Tests (RDTs) and Artemisinin-based combination therapies (ACTs), the emergence of pyrethroid resistance hindered malaria prevention efforts, prompting the WHO to recommend nets containing the synergist piperonyl butoxide (PBO) [53, 54]. Climate change and increased mobility are expected to reintroduce malaria to previously non-endemic areas [55].

The COVID-19 pandemic exacerbated stagnation in malaria control, particularly in Africa, where progress had already plateaued since 2015. Malaria incidence and mortality in Africa are off-track by 40% and 42%, respectively, compared to global targets [38]. Disruptions to malaria services in high-transmission African countries could double malaria deaths [31], with even moderate disruptions in diagnosis, treatment, bed-net distribution, and preventive measures having severe consequences [38]. In Africa, impact of pandemic-related disruptions varied by country. Uganda saw a slight decline in malaria testing due to supply chain issues, while caseloads remained like pre-COVID levels [60]. In contrast, Zimbabwe experienced rise in malaria-related cases and deaths between 2019 and 2020 [40]. However, differences in data sources complicate cross-country comparisons. To meet the 2030 malaria targets, countries need to tailor responses using locally generated data and secure more resources, as funding was diverted to COVID-19 response effort [36]. It is anticipated that early pandemic disruptions in malaria services will eventually result in a higher death toll.

4.5 Healthcare System Adaptability in the Post-epidemic Period

The pandemic has reshaped our understanding of healthcare system resilience, highlighted vulnerabilities, and sparked significant innovations in medical infrastructure and service delivery. This global health crisis exposed critical weaknesses in existing systems but also accelerated the adoption of new technologies that will influence healthcare for years. The interaction between technological advancements, healthcare delivery, and global preparedness has become a key feature of modern healthcare, challenging traditional approaches to patient care and medical infrastructure [56].

4.6 Socioeconomic Implications of Communicable Diseases

The pandemic has severely impacted low-income migrant workers, with many facing economic hardship. Disturbing images of migrant workers walking home during the lockdown sparked significant debate. Remittances from these workers play a crucial role in poverty reduction and economic development. In 2019, South Asia received $139 billion (₹1,04,2500 crores) in remittances from countries such as those in the Gulf. However, COVID-19 has caused a major decline, with India's remittances expected to drop by 23% in 2020, from $83 billion (₹6,20,000 crores) in 2019 to $64 billion (₹4,80,000 crores). The World Economic Forum reports that many migrants stranded abroad are forced to take low-paid jobs, live a poor life, and expenses restricted spending, which increases their exposure to COVID-19. Social life gets segregated on an economic basis. India's social fabric is built on strong emotional and economic interdependence within families and communities. Close physical interactions, such as living in crowded spaces and social gatherings, make social distancing difficult, even during the pandemic. Despite lockdown measures, crowding has been observed in religious places, during travel, and at liquor stores. While "vertical distancing" highlights existing inequalities, the "horizontal distancing" imposed by COVID-19 has worsened these disparities. The scale of the issue has made government schemes insufficient. 12.96 lakh metric tonnes of food grains were recently distributed by the Food Corporation of India as part of the Pradhan Mantri Garib Kalyan Anna Yojna (PMGKAY) to mitigate the effects of COVID-19 [57].

4.7 Research and Future Directions

At the outset of the COVID-19 pandemic, efforts to identify effective treatments focused on repurposing existing drugs with known therapeutic properties. Dexamethasone, a corticosteroid, emerged as a key candidate due to its potential benefits in treating severe cases [58]. While the independent COVID-19 Therapeutics Advisory Panel (UK-CTAP) made crucial recommendations on which medicines should be given priority for testing in national clinical trials, the New and Emerging Respiratory Virus Threats Advisory Group (NERVTAG) produced a list of intriguing choices [59]. As these trials progressed, the scope broadened to include novel therapies. One such addition was *Regeneron's* REGN-COV2, which is a monoclonal antibody cocktail specifically designed to target the virus and prevent severe disease progression. Traditional public health measures, such as isolating infected individuals and minimizing transmission through reduced contact, specifically prevent infectious diseases. However, innovative scientific tools have become increasingly significant. Advanced diagnostic technologies like loop-mediated isothermal amplification (LAMP), lateral flow devices (LFDs), and LamPORE are crucial for quickly identifying cases and preventing small clusters from evolving into widespread outbreaks. Vaccination is another critical element in managing the pandemic. For example, by late 2021, billions of doses had been administered worldwide, highlighting the potential for rapid vaccine deployment during a global crisis. Distribution must be equitable because not only a few people or places have to access a life-saving intervention at the right time but also in those vulnerable populations and low-resource settings it is a strategic plan for policymakers. Another strategy to hasten emergencies during health crises is to simplify and streamline procedures for the approval of vaccines and therapeutics. Always, a key area should be to maintain and strengthen robust supply chains avoiding interruptions in the availability of essential medical supplies.

Global collaboration during the pandemic has proven invaluable and highlights the importance of establishing international partnerships and policy frameworks that allow for rapid, coordinated response to future pandemics. Adaptively inclusive and efficient policies concern preparedness and resilience from threats of new infectious diseases (Fig. 4.2).

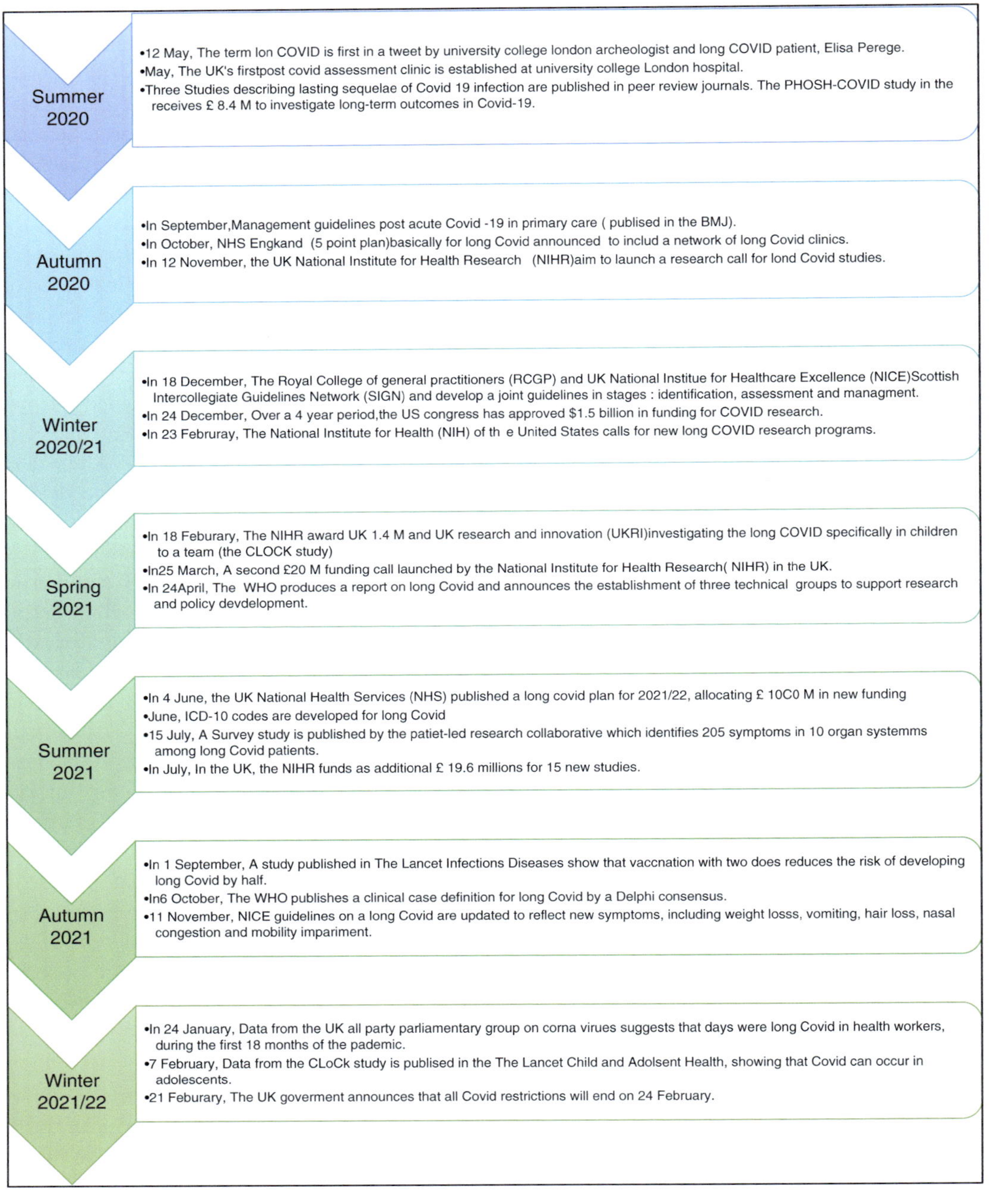

Fig. 4.2 International timeline-long COVID policy strategies [61]

4.8 Conclusion

The post-COVID-19 era underscores the complexity of communicable diseases as challenges that extend far beyond the medical realm. Interwoven social, economic, and technological factors play pivotal roles in shaping disease dynamics, demanding comprehensive, adaptive approaches for understanding and managing these challenges. Future strategies must focus on developing resilient and adaptive healthcare systems promptly capable of responding to evolving threats. Addressing social determinants of health—poverty, education, and access to basic services—remains critical for reducing vulnerabilities. Simultaneously, investment in technological innovations, including advanced diagnostics, telemedicine, and data-driven disease modelling, will enhance early detection and equitable healthcare delivery. Collaboration across borders will also be essential, as global research partnerships enable faster development of vaccines, treatments, and shared best practices.

References

1. Menges D, Ballouz T, Anagnostopoulos A, Aschmann HE, Domenghino A, Fehr JS, Puhan MA. Burden of post-COVID-19 syndrome and implications for healthcare service planning: a population-based cohort study. PLoS One. 2021;16(7):e0254523. https://doi.org/10.1371/journal.pone.0254523.
2. GBD 2019 Child and Adolescent Communicable Disease Collaborators. The unfinished agenda of communicable diseases among children and adolescents before the COVID-19 pandemic, 1990–2019: a systematic analysis of the Global Burden of Disease Study 2019. Lancet. 2023;402(10398):313–35.
3. The Global Fund. TB testing in 2020 dropped drastically due to COVID-19. 2021. Accessed 14th Jan 2022 from https://www.theglobalfund.org/en/news/2021-03-24-tb-testing-in-2020-dropped-drastically-due-to--covid-19/
4. Hay SI. Maintaining progress for the most beautiful chart in the world. Int Health. 2019;11(5):344–8.
5. Cioboata R, Biciusca V, Olteanu M, Vasile CM. COVID-19 and tuberculosis: unveiling the dual threat and shared solutions perspective. J Clin Med. 2023;12(14):4784. https://doi.org/10.3390/jcm12144784 6.
6. Afum T, Asare P, Asante-Poku A, et al. Diagnosis of tuberculosis among COVID-19 suspected cases in Ghana. PLoS One. 2021;16(12):e0261849. https://doi.org/10.1371/journal.pone.0261849. Published 2021 Dec 28.
7. Mendez-Brito A, El Bcheraoui C, Pozo-Martin F. Systematic review of empirical studies comparing the effectiveness of non-pharmaceutical interventions against COVID-19. J Infect. 2021;83(3):281–93. https://doi.org/10.1016/j.jinf.2021.06.018.
8. Juneau CE, Pueyo T, Bell M, Gee G, Collazzo P, Potvin L. Lessons from past pandemics: a systematic review of evidence-based, cost-effective interventions to suppress COVID-19. Syst Rev. 2022;11(1):90. https://doi.org/10.1186/s13643-022-01958-9.
9. Ritchie H, Mathieu E, Rodes-Guirao L, Appel C, Giattino C, Ortiz-Ospina E, Hasell J, Macdonald B, Beltekian D, Roser M. Coronavirus pandemic (COVID-19). Our world in data. 2022. http://ourworldindata.org/coronavirus. Accessed 2022.
10. Ayouni I, Maatoug J, Dhouib W, Zammit N, Fredj SB, Ghammam R, Ghannem H. Effective public health measures to mitigate the spread of COVID-19: a systematic review. BMC Public Health. 2021;21(1):1015. https://doi.org/10.1186/s12889-021-11111-1.
11. Centers for Disease Control and Prevention. Communicable disease surveillance report. Atlanta: CDC Press; 2021.
12. Anderson RM, et al. How will country-based mitigation measures influence the course of COVID-19? Lancet. 2020;395(10228):931–4.
13. The Global Fund. Strategic initiatives update: CRG 2022-06. 2022. Retrieved from https://www.theglobalfund.org
14. White, et al. The global decline in disease notifications may have been exacerbated by underreporting, changes in healthcare-seeking behaviour, and decreased access to healthcare. 2021.
15. Centers for Disease Control and Prevention. STD facts—congenital syphilis. 2020. https://www.cdc.gov/std/syphilis/stdfact-congenital-syphilis.htm. Accessed 3 June 2020.
16. Kunz JM, Hannapel E, Vander Kelen P, Hils J, Hoover ER, Edens C. Effects of the COVID-19 pandemic on *Legionella* water management program performance across a United States lodging organization. Int J Environ Res Public Health. 2023;20(19):6885. https://doi.org/10.3390/ijerph20196885.
17. Olsen SJ, Azziz-Baumgartner E, Budd AP, Brammer L, Sullivan S, Pineda RF, Cohen C, Fry AM. Decreased influenza activity during the COVID-19 pandemic-United States, Australia, Chile, and South Africa, 2020. Am J Transplant. 2020;20(12):3681–5. https://doi.org/10.1111/ajt.16381.
18. Madhav N, Oppenheim B, Gallivan M, et al. Chapter 17: pandemics: risks, impacts, and mitigation. In: Jamison DT, Gelband H, Horton S, et al., editors. Disease control priorities: improving health and reducing poverty. 3rd ed. Washington, DC:

The International Bank for Reconstruction and Development/The World Bank; 2017. Available from: https://www.ncbi.nlm.nih.gov/books/NBK525302/. https://doi.org/10.1596/978-1-4648-0527-1_ch17.
19. Xue B, Wu Y, Zheng X, Xue Y, Dong F, Xiao S, et al. Effects of information sources on public preventive behaviours in health emergencies: evidence from a digital epidemiologic study during the COVID-19 pandemic. Front Public Health. 2022;10:981649.
20. European Centre for Disease Prevention and Control. Digital technologies for the surveillance, prevention, and control of infectious diseases: a scoping review of the research literature 2015–2019 (ECDC Technical Report). 2020. https://www.ecdc.europa.eu/sites/default/files/documents/Digital-technologies-for-infectious-diseases.pdf
21. Coalition for Epidemic Preparedness Innovations, Gavi, & World Health Organization. COVAX: ensuring rapid and equitable access to COVID-19 vaccines. 2020.
22. Ramazzotti D, Maspero D, Angaroni F, Spinelli S, Antoniotti M, Piazza R, Graudenzi A. Early detection and improved genomic surveillance of SARS-CoV-2 variants from deep sequencing data. iScience. 2022;25(6):104487. https://doi.org/10.1016/j.isci.2022.104487.
23. Cheng Y, Ji C, Zhou HY, Zheng H, Wu A. Web resources for SARS-CoV-2 genomic database, annotation, analysis and variant tracking. Viruses. 2023;15(5):1158. https://doi.org/10.3390/v15051158.
24. Satam H, Joshi K, Mangrolia U, Waghoo S, Zaidi G, Rawool S, Thakare RP, Banday S, Mishra AK, Das G, Malonia SK. Next-generation sequencing technology: current trends and advancements. Biology. 2023;12(7):997. https://doi.org/10.3390/biology12070997.
25. John G, Sahajpal NS, Mondal AK, Ananth S, Williams C, Chaubey A, Rojiani AM, Kolhe R. Next-generation sequencing (NGS) in COVID-19: a tool for SARS-CoV-2 diagnosis, monitoring new strains and phylodynamic modeling in molecular epidemiology. Curr Issues Mol Biol. 2021;43(2):845–67. https://doi.org/10.3390/cimb43020061.
26. World Health Organization. Maintaining essential health services during COVID-19 pandemic. World Health Organization; 2022.
27. Haldane V, et al. Health system resilience in managing the COVID-19 pandemic: lessons from Singapore. BMJ Glob Health. 2021;6(1):e004216.
28. Remuzzi A, Remuzzi G. COVID-19 and Italy: what next? Lancet. 2020;395(10231):1225–8.
29. Shapira G, Ahmed T, Drouard SHP, Fernandez PA, et al. Disruptions in maternal and child health service utilization during COVID-19: analysis from eight sub-Saharan African countries. Health Policy Plan. 2021;36(7):1140–51. https://doi.org/10.1093/heapol/czab064.
30. Holtzman CW, Godfrey C, Ismail L, et al. PEPFAR's role in protecting and leveraging HIV services in the COVID-19 response in Africa. Curr HIV/AIDS Rep. 2022:1–11. https://doi.org/10.1007/s11904-021-00587-6.
31. Weiss DJ, Bertozzi-Villa A, Rumisha SF, et al. Indirect effects of the COVID-19 pandemic on malaria intervention coverage, morbidity, and mortality in Africa: a geospatial modelling analysis. Lancet Infect Dis. 2021;21(1):59–69. https://doi.org/10.1016/S1473-3099(20)30700-3.
32. Nachega JB, Sam-Agudu NA, Machekano RN, et al. Assessment of clinical out-comes among children and adolescents hospitalized with COVID-19 in 6 Sub-Saharan African countries. JAMA Pediatr. 2022;176(3):e216436. https://doi.org/10.1001/jamapediatrics.2021.6436. Published online January 19.
33. Gondwe G. Assessing the impact of COVID-19 on Africa's economic development. UNCTAD/ALDC/MISC/2020/3. Accessed 4th Feb 2022 at https://unctad.org/system/files/officialdocument/aldcmisc2020d3_en.pdf
34. Inzaule SC, Ondoa P, Loembe MM, Tebeje YK, Ouma AEO, Nkengasong JN. COVID-19 and indirect health implications in Africa: impact, mitigation measures, and lessons learned for improved disease control. PLoS Med. 2021;18(6) https://doi.org/10.1371/journal.pmed.1003666. PMID: 34161318; PMCID: PMC8266084.
35. Loembé MM, Nkengasong JN. COVID-19 vaccine access in Africa: global distribution, vaccine platforms, and challenges ahead. Immunity. 2021;54(7):1353–62. https://doi.org/10.1016/j.immuni.2021.06.017. PMID: 34260880; PMCID: PMC8276532.
36. World Health Organisation (WHO). WHO calls for reinvigorated action to fight malaria. 2020. Accessed 14th Jan 2022 at https://www.who.int/news/item/30-11-2020-who-calls-for-reinvigorated-action-to-fight--malaria
37. Maharana A, Amutorine M, Sengeh MD, Nsoesie EO. COVID-19 and beyond: use of digital technology for pandemic response in Africa. Sci Afr. 2021;14:e01041. https://doi.org/10.1016/j.sciaf.2021.e01041.
38. World Health Organisation (WHO). World malaria report 2021. Geneva: World Health Organization; 2021. Licence: CC BY-NC-SA 3.0 IGO. Accessed 14th Jan 2022 at https://www.who.int/teams/global-malaria-programme/reports/world-malaria-report-2021
39. Hoke T, Bateganya M, Toyo O, et al. How home delivery of antiretroviral drugs ensured uninterrupted HIV treatment during COVID-19: experiences from Indonesia, Laos, Nepal, and Nigeria. Glob Health Sci Pract. 2021;9(4):978–89. https://doi.org/10.9745/GHSP-D-21-00168. PMID: 34933991; PMCID: PMC8691873.
40. Gavi S, Tapera O, Mberikunashe J, Kanyangarara M. Malaria incidence and mortality in Zimbabwe during the COVID-19 pandemic: analysis of routine surveillance data. Malar J. 2021;20(1):233. https://doi.org/10.1186/s12936-021-03770-7.

41. Ung L, Stothard JR, Phalkey R, et al. Towards global control of parasitic diseases in the Covid-19 era: one health and the future of multisectoral global health governance. Adv Parasitol. 2021;114:1–26. https://doi.org/10.1016/bs.apar.2021.08.007. Epub 2021 Oct 11. PMID: 34696842; PMCID: PMC8503781.
42. Zumla A, Dar O, Kock R, et al. Taking forward a 'One Health' approach for turning the tide against the Middle East respiratory syndrome coronavirus and other zoonotic pathogens with epidemic potential. Int J Infect Dis. 2016;47:5–9.
43. Kapata N, Ihekweazu C, Ntoumi F, et al. Is Africa prepared for tackling the COVID-19 (SARS-CoV-2) epidemic. Lessons from past outbreaks, ongoing pan-African public health efforts, and implications for the future. Int J Infect Dis. 2020;93:233–6.
44. McQuaid CF, Vassall A, Cohen T, Fiekert K, White RG. The impact of COVID-19 on TB: a review of the data. Int J Tuberc Lung Dis. 2021;25(6):436–46. https://doi.org/10.5588/ijtld.21.0148.
45. World Health Organisation (WHO). Global tuberculosis report 2021. Geneva: World Health Organization; 2021. Licence: CC BY-NC-SA 3.0 IGO. Accessed 14 th Jan 2022 at https://www.who.int/teams/global-tuberculosis-programme/tb-reports/global-tuberculosis-report-2021
46. Tessema GA, Kinfu Y, Dachew BA, et al. The COVID-19 pandemic and healthcare systems in Africa: a scoping review of preparedness, impact, and response. BMJ Glob Health. 2021;6(12) https://doi.org/10.1136/bmjgh-2021-007179.
47. Tamuzi JL, Ayele BT, Shumba CS, et al. Implications of COVID-19 in high bur- den countries for HIV/TB: a systematic review of evidence. BMC Infect Dis. 2020;20:744. https://doi.org/10.1186/s12879-020-05450-4.
48. Saunders MJ, Evans CA. COVID-19, tuberculosis, and poverty: preventing a perfect storm. Eur Respir J. 2020;56(1):2001348. https://doi.org/10.1183/13993003.01348-2020. Published 2020 Jul 9.
49. Matos R, Fonseca KL, Mereiter S, et al. Mycobacterium tuberculosis infection up-regulates Sialyl Lewis X expression in the lung epithelium. Microorganisms. 2021;9:99. https://doi.org/10.3390/microorganisms9010099.
50. Jewell BL, Mudimu E, Stover J, et al. Potential effects of disruption to HIV programmes in sub-Saharan Africa caused by COVID-19: results from multiple mathematical models. Lancet HIV. 2020;7(9):e629–40. https://doi.org/10.1016/S2352--3018(20)30211-3. Epub 2020 Aug 6. PMID: 32771089; PMCID: PMC7482434.
51. Stover J, Kelly SL, Mudimu E, et al. The risks and benefits of providing HIV services during the COVID-19 pandemic. PLoS One. 2021;16(12) https://doi.org/10.1371/journal.pone.0260820. PMID: 34941876; PMCID: PMC8699979.
52. Dorward J, Khubone T, Gate K, et al. The impact of the COVID-19 lockdown on HIV care in 65 South African primary care clinics: an interrupted time series analysis. Lancet HIV. 2021;8(3):e158–65. https://doi.org/10.1016/S2352-3018(20)30359-3. Epub 2021 Feb 4. PMID: 33549166; PMCID: PMC8011055.
53. Churcher TS, Lissenden N, Griffin JT, et al. The impact of pyrethroid resistance on the efficacy and effectiveness of bednets for malaria control in Africa. elife. 2016;5:e16090. https://doi.org/10.7554/eLife.16090. PMID: 27547988; PMCID: PMC5025277.
54. Lindsay SW, Thomas MB, Kleinschmidt I. Threats to the effectiveness of insecticide-treated bednets for malaria control: thinking beyond insecticide resistance. Lancet Glob Health. 2021;9(9):e1325–31. https://doi.org/10.1016/S2214-109X(21)00216-3. Epub 2021 Jun 30. PMID: 34216565.
55. Cella W, Baia-da-Silva DC, Melo GC, et al. Do climate changes alter the distribution and transmission of malaria? Evidence assessment and recommendations for future studies. Rev Soc Bras Med Trop. 2019;52:e20190308. https://doi.org/10.1590/0037--8682-0308-2019. PMID: 31800921.
56. Ranzani Rigotti A, Mara Zamarioli C, do Prado PR, Helena Pereira F, Gimenes FRE. Resilience of healthcare systems in the face of COVID-19: an experience report. Rev Esc Enferm USP. 2022;56:e20210210. https://doi.org/10.1590/1980-220X-REEUSP-2021-0210en.
57. Gopalan HS, Misra A. COVID-19 pandemic and challenges for socio-economic issues, healthcare, and National Health Programs in India. Diabetes Metab Syndr. 2020;14(5):757–9. https://doi.org/10.1016/j.dsx.2020.05.041.
58. DHSC. World first coronavirus treatment approved for NHS use by government. 2020. https://www.gov.uk/government/news/world-first-coronavirus-treatment-approved-for-nhs-use-by-government
59. DHSC. Guidance: making a proposal for COVID-19 therapeutics clinical trials. 2020. https://www.gov.uk/government/publications/covid-19-treatments-making-a-proposal-for-clinical-trials/guidance-making-a-proposal-for-covid-19-therapeutics-clinical-trials
60. Chanda-Kapata P, Ntoumi F, Kapata N, Lungu P, Mucheleng'anga LA, Chakaya J, Tembo J, Himwaze C, Ansumana R, Asogun D, Mfinanga S, Nyasulu P, Mwaba P, Yeboah-Manu D, Zumla A, Nachega JB. Tuberculosis, HIV/AIDS, and malaria health services in sub-Saharan Africa—a situation analysis of the disruptions and impact of the COVID-19 pandemic. Int J Infect Dis. 2022b;124:S41–6. https://doi.org/10.1016/j.ijid.2022.03.033.
61. Prashar J. Long covid: conceptualizing the challenges for public health. J Public Health. 2023;45(3):771–9. https://doi.org/10.1093/pubmed/fdac153.
62. Working for global equitable access to COVID-19 vaccines. Available online: https://www.who.int/initiatives/act-accelerator/covax. Accessed on 15 June 2022.
63. Lancet N. COVID-19 in Latin America—emergency and opportunity. Lancet. 2021;398(10295):93. https://doi.org/10.1016/s0140-6736(21)01551-8.

COVID-19 Global Vaccination Drive

5

Saurabh Mishra, Hamid Ashraf, Jamal Ahmad,
Goo-Bo Jeong, and Alok Raghav

5.1 Introduction

Since late 2019, COVID-19 has dominated global discussions as the third major coronavirus pandemic, following MERS-CoV (2012) and SARS-CoV (2002). SARS-CoV-2, the virus causing COVID-19, shares genetic similarities with SARS-CoV BJ01 (75%, GenBank: AY278488.2), MERS-CoV HCoV-EMC (55%, GenBank: MH454272.1), and RaTG13 (96%, GenBank: MN996532.1), highlighting its evolutionary link within the coronaviridae family. Phylogenetic analyses have clarified its origins, emphasizing its close relationship with earlier coronavirus strains [1–4]. SARS-CoV-2 infection presents a wide range of clinical outcomes, from no acute symptoms to chronic pneumonia, fatal acute respiratory distress syndrome, mild flu-like illness, severe pneumonia, and mild flu-like illness. Even asymptomatic cases are critical, as they have significantly contributed to the virus's rapid spread globally. The WHO implemented measures to prevent the pandemic, including boosting immunity, encouraging personal hygiene, and isolating suspected (undocumented or infected) and confirmed (confirmed) cases. Despite these efforts, the virus's impact was devastating, with nearly all nations struggling to manage its effects. The current coronavirus outburst was declared a pandemic by the World Health Organization in March 2020. According to the WHO database (https://covid19.who.int), 670,000,000 cases and more than 6,700,000 deaths have been reported since the pandemic began, with the United States accounting for the largest share of infections and deaths, respectively. Globally, COVID-19 has infected 1900 persons every week since December 2022. The COVID-19 mortality rate has been 12,000 per week since the year began. Some people may be concerned about this report. Since the beginning of 2023, the number of deaths has escalated to 5900 in the United States and 2700 in Europe every week. There were rumours that the increased focus on creating more potent drugs was one of the factors delaying the release of COVID-19 vaccinations. Even yet, the world's population would gain more from the earlier delivery of a less effective vaccine than from postponing the immunization campaign to wait for a more effective one [5]. Giving a somewhat effective vaccination before

S. Mishra
Department of Life Sciences and Biotechnology,
Chhatrapati Shahu Ji Maharaj University,
Kanpur, Uttar Pradesh, India

H. Ashraf · J. Ahmad
Rajiv Gandhi Centre for Diabetes and Endocrinology,
J.N Medical College, Aligarh Muslim University,
Aligarh, Uttar Pradesh, India

G.-B. Jeong · A. Raghav (✉)
Department of Anatomy and Cell Biology,
Gachon University Gil Medical Center,
Incheon, Korea (Republic of)
e-mail: gbjeong@gachon.ac.kr; alokalig@gachon.ac.kr

A. Raghav et al. (eds.), *COVID-19 Impact on Economy, Environment and Healthcare*,
https://doi.org/10.1007/978-981-95-1196-9_5

winter, for instance, may have considerably decreased last year's incidence peaks, resulting in lower morbidity and death and less strain on medical staff, since the prevalence of viral diseases rises during the winter months [6, 7]. The United States had enough doses to follow the recommended schedule of 3–4 weeks, but by the end of 2020, the UK was the first nation to recommend a 12-week meantime between the two doses alternatively to the previously recommended 3–4 week interval for the AstraZeneca and Pfizer-BioNTech vaccines [8]. The delay was justified by the need to safeguard the greatest number of individuals who were at risk in the shortest amount of time. It was believed that this approach would lower COVID-19-related morbidity, mortality, and hospitalizations even in cases of weakened immunity. Delaying the second dosage can also help to prolong the duration of the vaccine-induced immune response, which is the best method to deal with the vaccines' declining immunity over time [9]. Numerous studies have set the recommended dosage parameters for various vaccine types to administer the first and second coronavirus vaccination dosages. This alternative vaccination approach is justified by the need to offer supplementary protection [10]. Compared to a single-dose or two-dose homologous regimen, a two-dose heterologous coronavirus immunization schedule has been shown to generate stronger antibody responses. Furthermore, this can contribute to wider population coverage in circumstances where the availability of vaccines is restricted. Additionally, switching to a different vaccination type would be a sensible choice if questions were raised regarding the original vaccine's safety or effectiveness [11].

5.2 COVID-19 Impact on Major Vital Organs

Blood and neurological system connect all the body's major organs. If the infection occurs in any organ, it is harmful to other vital organs. Recent research, however, has revealed that SARS-CoV-2 has also been correlated to negative impacts on other vital organs. We are aware that the virus primarily affects the lungs by generating acute respiratory distress syndrome (ARDS), which results in diffuse alveolar destruction: the heart, kidney, spleen, big intestine, and brain [12–14]. According to the research that has been done so far old age people with some health issues are severely affected [15]. This virus-causing disease is highly contagious and has infected over 212 million people and claimed over 4.4 million people cause of death. Over 38.5 million people are infected in the United States alone, and as of August 23, 2021, there have been approximately 0.645 million fatalities, drastically increasing the number of people day by day [20]. During the post-COVID period, scientists worked hard to resolve the problem of this new coronavirus infection. To fully understand the reasons for negative outcomes in coronavirus patients, examining the underlying cause of tissue injury in these patients is significant [16]. Here, scientists did studies linked to the infection pathway, transmission mechanism, and human organ malfunction by SARS-CoV-2. The dangerously affected organ is the lung [17, 18]. In Wuhan, China, 62 individuals with confirmed COVID-19 pneumonia were retrospectively evaluated. COVID-19 pneumonia patients who underwent computerized tomographic testing showed involvement of the interstitium and lung parenchyma [20]. Injuries in the lower, middle, and posterior lung areas showed a distinctive multifocal distribution. The increased levels of high-sensitivity C-reactive protein (hs-CRP), prime cause of heart strokes, and a low lymphocyte count were seen [18]. When pneumonia worsens, critical illness patients develop ARDS [19, 21]. With a higher death rate, acute respiratory distress syndrome is the lethal adverse consequence of SARS-CoV-2. 32 out of the 52 severely infected patients in the study died, and 35 of them had ARDS [22]. The evidence indicates that SARS-CoV-2 and other coronaviruses infect sites in the human body through the nasal and oropharyngeal mucosa, and some of them eventually accumulate in the lungs. Entering the lung cells, SARS-CoV-2 destroys the cilia, which are hair-like projections that move to keep the airways clean of mucus or debris and begin to replicate. Lung cells eventually die and shed, allowing debris to enter, and impairing the body's ability to keep substances out of the trachea and lungs. The patient has a dry cough and trouble breathing because of inflammation being set off, fluid leaking into the alveolar gaps, and no gas exchange taking place. The next sensitive organ is the heart affected by this new strain of coronavirus. The most recent COVID-19 strain has been reported to cause cardiac problems, in contrast to the seven (7) known strains of human coronaviruses, which are known to impact the respiratory system but not heart [20]. One of the most prevalent diseases associated with COVID-19 is cardiovascular disease, according to recent data from Italy, Wuhan, and the United States [23–25]. Additionally, the mortality rate for COVID-19 patients with cardiovascular disease is 10.5% higher than that of individuals with hypertension (6%), diabetes (7.3%), and chronic respiratory disease (6.3%) [26]. However, the exact mechanisms causing the heart damage caused by SARS-CoV-2 infection are unknown. The effects of SARS-CoV-2 on the heart can occur in two ways. ACE2 receptors are widely distributed in the heart and lungs, and the virus may first enter the heart directly and harm it through these receptors [27, 28]. According to some reports, liver impairment is a non-pulmonary symptom of the SARS-CoV-2 virus. COVID-19 patients exhibit a range of changes in liver-associated enzyme levels, including aspartate aminotransferase (AST), total bilirubin (TBIL), alanine aminotransferase (ALT), gamma-glutamyl transferase (GGT), alkaline phosphatase (ALP), and albumin [29–31]. Levels of AST, ALT, ALP, albumin, and TBI were examined in a study of 1827 individuals with confirmed COVID-19 at the Yale-New Haven Health System in the United States between March and April 2020. ALT 41.6%, AST 66.9%, ALP 13.5%, and TBIL 4.3% were abnormal liver test results at admission, while AST 83.4%, ALT 61.6%, ALP 22.7%, and TBIL 16.1% were abnormal liver test results during maximum hospitalization [30]. One (1·8%) and thirty (54%) of 56 COVID 19 patients in another study had increased ALP and GGT levels while they were in the hospital [32]. Numerous investigations have demonstrated various gastrointestinal symptoms brought on by SARS-CoV-2 infection. According to a 60 studies meta-analysis of 4243 participants, the combined rate of GI symptoms (17.6%). Anorexia accounted 26.8% of the most common documented symptoms in this investigation, followed by acute diarrhoea (12.5%), stomach discomfort (9.2%), nausea and vomiting (10.2%) [33]. Regarding the infection pathway, it has been documented that the human kidneys and bladders express the ACE2 gene. 9.2% reported gastrointestinal pain and 10.2% reported vomiting [33]. According to the findings, the respiratory, hepatic, and digestive systems, as well as the urinary system, are possible entry points for 2019-SARCoV infection. So, for treatment purposes, vaccine development strategies have been initiated by many scientists. The long-awaited permits of the first COVID-19 vaccine in December 2020 were intended to help alleviate the pandemic's impacts. There are now ten vaccines that the WHO recommends, and many more are being developed [37]. Lack of access to vaccines may make it impossible for countries to develop and carry out vaccination campaigns that increase public knowledge and vaccination compliance [34]. The protection of public health depends on widespread COVID-19 vaccinations, which has been proven to be very successful in preventing severe COVID-19 infections and death from the disease (Public Health Ontario 2021). Also, if the transmission of COVID-19 is not controlled, novel and vaccine-resistant variants of SARS-CoV-2 are more probable to emerge [35]. However, the aim of universal vaccinations is seriously threatened by vaccine reluctance or hesitancy (Thunstrom 2020) [36]. The development of vaccines against the Middle East respiratory syndrome coronavirus (MERS-CoV) and SARS-CoV has advanced, according to several research [7, 8, 38, 39]. The development of the now commercialized SARS-CoV-2 vaccines was sped up in part by preclinical data on candidate vaccines, which would also serve as platforms for the vaccines' eventual broad use. The

World Health Organization (WHO) classifies COVID-19 vaccines as vector, protein component, DNA, RNA, inactivated, live attenuated, and virus-like particle vaccinations after they have been analysed or approved for clinical investigations. When developing vaccines, animal studies are essential, comprising defining the appropriate dosage, choosing the injection schedule, and assessing the safety and protective effectiveness. Small animals are tested to generate vaccines, and biological and immunological research is based on animals, particularly rats [34, 35].

5.3 Pre-existing COVID-19 Vaccine Platforms

WHO data published on March 28, 2022, shows that 196 vaccinations are going through preclinical testing and 153 vaccines have been authorized for clinical trials. These vaccines mainly consist of the following: DNA vaccines (11%), RNA vaccines (18%), live attenuated vaccines (1%), viral vector vaccines (both non-replication and replication; 17% of the total), protein subunit vaccines (34%), and VLP vaccines (4%) mRNA as shown in Table 5.1.

Table 5.1 List of vaccines [43]

Sr. no.	Manufacturer	Name of vaccine	Phase of clinical trial	Vaccine platform
1.	Meissa Vaccines, Inc.	A live attenuated vaccine termed MV-014-212 expresses the spike (S) protein of SARS-CoV-2.	Phase 1	Live attenuated virus
2.	Radboud University	ABNCoV2 capsid virus-like particle (cVLP) +/– adjuvant MF59	Phase 1	A particle that resembles a virus
3.	National Research Centre, Egypt	CoviVax®	Phase 1	Whole virus inactivated
4.	Kocak Farma	Inactivated adjuvant Kocak COVID-19 vaccination	Phase1	Inactivated
5.	VBI Vaccines Inc.	The VBI-2902a. An aluminium phosphate adjuvant with SARS-CoV-2 spike (S) glycoprotein enveloped virus-like particle (eVLP).	Phase 1/2	Virus-like particle
6.	The national agency of Turkey (Scientific and Technological Research Council of Turkey)	VLP for SARS-CoV-2 Immunization Vaccine-Alpha, Vaccine-Wuhan, and Vaccine-Wuhan + Alpha variant	Phase 2	Virus-like particle
7.	Chumakov Federal Scientific Centre for Development of Immune-and-Biological Products	CoviVac®	Phase 1/2	Inactivated
8.	Vaccibody AS	The DNA plasmid vaccine VB10.2210 encodes several immunogenic and conserved cell epitopes that span several antigens throughout the genome of SARS-CoV-2.	Phase1/2	Vaccination based on DNA
9.	University of Munich (Ludwig-Maximilians)	MVA-SARS-2-S	Phase1	The non-replicating viral vector
10.	National Research Centre, Egypt	CoviVax®	Phase 1	Inactivated whole virus inactivated

(continued)

Table 5.1 (continued)

Sr. no.	Manufacturer	Name of vaccine	Phase of clinical trial	Vaccine platform
11.	Serum Institute of India + Accelagen Pty + Spy Biotech	RBD SARS-CoV-2 HBsAg VLP vaccine	Phase1/2	Virus-like particle
12.	The Scientific and Technological Research Council of Turkey (TUBITAK)	Adjuvanted inactivated vaccine against SARS-CoV-2	Phase 1	Inactivated
13.	Osve Pharmaceutical Company	Osvid-19®	Phase1	Inactivated
14.	"Aivita Biomedical, Inc.; "National Institute of Health Research and Development; Ministry of Health, Republic of Indonesia	Vaccination with dendritic cells, AV-COVID-19. A vaccine that uses autologous dendritic cells loaded with SARS-CoV-2 antigens, either with or without GM-CSF	Phase2	Viral vector (Replicating) + APC
15.	The Government Pharmaceutical Organization (GPO): PATH, Dynavax	Inactivated (NDV-based) Chimeric vaccine with or without adjuvant CpG 1018	Phase 1/2	Inactivated + CpG1018
16.	Kentucky Bioprocessing Inc.	KBP-COVID-19(RBD-based)	Phase1/2	Protein subunit
17.	Vaxart	VXA-CoV2–1Ad5 adjuvanted Oral Vaccine platform	Phase2	The non-replicating viral vector
18.	ReiThera + Leukocare + Univer cells	Simian Adenovirus (GRAd) with replication defects that encode S is known as GRAd-COV2.	Phase2/3	The non-replicating viral vector
19.	Arcturus Therapeutics	ARCT-021	Phase2	Vaccination based on RNA
20.	Shifa Pharmed Industrial Co	Coviran-Barkat®	Phase2/3	Inactivated
21.	Genexine Consortium	GX-19N	Phase2/3	Vaccination based on DNA
22.	AnGes + TakaraBio + Osaka University	AG0301-COVID19	Phase2/3	Vaccination based on DNA
23.	KM Biologic Co Ltd	Inactivated COVID-19 vaccine	Phase2/3	Inactivated + alum
24.	Medicago Inc.	COVID-19 Coronavirus-Like Particle (CoVLP) MT-2766 Alternative Name: AS03 adjuvant, CoVLP.	Phase3	Virus-like particle
25.	Erciyes University and the Health Institute of Turkey (TUSEB)	TURKOVAC®	Phase3	Inactivated
26.	Codagenix/Serum Institute of India	COVI-VAC	Phase3	Live attenuated virus
27.	Research Institute for Biological Safety Problems, Rep of Kazakhstan	QazCOVID-in®	Phase 3	Inactivated

(continued)

Table 5.1 (continued)

Sr. no.	Manufacturer	Name of vaccine	Phase of clinical trial	Vaccine platform
28.	Sanofi Pasteur + GSK Phase3	VAT00008: SARS-CoV 2S protein with adjuvant (1) CoV2 preS dTM monovalent D614 antigen, (2) Bivalent (2-antigen) vaccine comprising spike protein of D614 and spike protein of the SARS-CoV-2Beta	Phase3	Protein subunit
29.	Valneva, National Institution of Health Research UK	VLA2001®	Phase 3	Inactivated
30.	Zydus Cadila	nCov vaccine	Phase3	Vaccination based on DNA
31.	Inovio Pharmaceuticals + International Vaccine Institute + Advaccine (Suzhou) Biopharmaceutical Co., Ltd	INO-4800 + electroporation	Phase3	Vaccination based on DNA
32.	Institute of Medical Biotechnology Chinese Academy of Medical Sciences (IMBCAMS), China	SARS-CoV-2 Vaccine, Inactivated (Vero Cell)	Phase3	Inactivated
33.	Gamaleya Research Institute; Health Ministry of the Russian Federation	Gam-COVID-Vac Adeno-based (rAd26-S + rAd5-S) Sputnik V COVID-19 vaccine	Phase3	The non-replicating viral vector
34.	CureVac AG	CVnCoV Vaccine	Phase3	Vaccine based on RNA
35.	Novavax	Full-length recombinant SARS CoV-2 glycoprotein nanoparticle vaccine adjuvanted with Matrix M (SARS-CoV-2rS/Matrix M1-Adjuvant) NVX-CoV2373	Phase3	Subunit of protein
36.	Bharat Biotech International Ltd	SARS-CoV-2Vaccine, Inactivated (Vero Cell)/ COVAXIN®	Phase3	Whole-virion inactivated
37.	Anhui Zhifei Longcom Biopharmaceutical + Institute of Microbiology, Chinese Academy of Sciences Zhongyianke Biotech Liaoning Maokangyuan Biotech Academy of Military Medical Sciences	Recombinant SARS CoV-2 vaccine (CHO Cell)	Phase 3	Subunit of protein
38.	Moderna + National Institute of Allergy and Infectious Diseases (NIAID)	mRNA-1273 Spikevax	Phase4	Vaccine based on RNA
39.	Astra Zeneca and the University of Oxford	ChAdOx1 S—(AZD1222) "Covishield" Vaxzevria	Phase4	The non-replicating viral vector
40.	Sinovac Research and Develop company Ltd	CoronaVac®: Inactivated SARS-CoV-2 vaccine (Vero Cell)	Phase4	Inactivated, produced in vero cells

(continued)

Table 5.1 (continued)

Sr. no.	Manufacturer	Name of vaccine	Phase of clinical trial	Vaccine platform
41.	CanSino Biological Inc. (Beijing Institute of Biotechnology)	New Coronavirus Recombinant Vaccine Ad5-nCoV (Adenovirus Type 5 Vector)	Phase4	The non-replicating viral vector
42.	Sinopharm BIBP (Beijing Institute of Biological Products Co., Ltd.)	Vero Cell's inactivated SARS-CoV-2 vaccine. COVID-19 BBIBP vaccination.	Phase4	Derived in Vero cells and inactivated
43.	Janssen Pharmaceutical Johnson & Johnson	Ad26.COV2-S	Phase4	The non-replicating viral vector
44.	Sinopharm WIBP (Wuhan Institute of Biological Products Co., Ltd.)	Inactivated SARS-CoV-2 Vaccine (Vero Cell)	Phase4	Derived in Vero cells and inactivated

5.4 Vaccines Safety

The currently licensed COVID-19 vaccines include many adverse effects, e.g., fever, exhaustion, injection site pain, headache, and nausea, despite being safe in clinical trials. Some participants experienced difficulties as the vaccine program proceeded on, and several individuals passed away from cardiovascular conditions such as arteriosclerosis. Additionally, in the mRNA-1273 and BNT162b2 vaccines' Phase III clinical studies, cardiac arrest. Most of the possible COVID-19 vaccine-related problems fall into the following categories. (1) disorders of coagulation, including thrombocytopenia; (2) heart conditions like myocarditis; (3) immune conditions like allergic reactions, autoimmune thyroid, and autoimmune hepatitis conditions; (4) the nervous system diseases, including facial paralysis and neurological disorders, and (5) diseases related to lymphatic system; and (6) other conditions, including Rowell's syndrome, 161 macular rash, and 162 chilblain-like lesions.

5.5 COVID-19 Vaccination's Effects on Different Populations

The COVID-19 vaccination primarily works by triggering memory cells and neutralizing antibodies. A history of allergies or malignancies, as well as innate immune disorders like autoimmune rheumatism, may make the COVID-19 immunization problematic for certain people. Pregnant and elderly women are additional causes for concern. Due to their compromised immune systems, older people may not receive the recommended protective impact from vaccines when compared to adults. Pregnancy-related side effects from the COVID-19 vaccination include fetal deformity, early birth, and abortion in women who are pregnant. According to earlier research, pregnant women who contract SARS-CoV-2 have a greater probability than non-pregnant women to experience problems such as lung damage, diabetes, and cardiovascular illnesses. After COVID vaccination, pregnant women suffer from adverse consequences such as abortion or foetal malformation. The important adverse effect was injection site pain. 9.4% of the pregnant women experienced an early birth, 86.1% had a normal pregnancy, and 13.9% experienced pregnancy loss following the mRNA vaccine. Premature birth and pregnancy loss are potential outcomes, but they are both not probable, and vaccination advantages outweigh the dangers. Furthermore, percentage of systemic or local adverse reactions in older women was comparable to that in pregnant women, suggesting pregnancy-related physiological changes had little effect on the incidence of adverse events. Pregnant women are still at a lower risk than non-pregnant women for adverse effects following vaccination. Within a recommended range, the COVID-19 vaccination protects both women and babies at same time, somewhat lowering the risk of postpartum fatal infection of SARS-CoV-2. Thus, it is recommended that pregnant women willingly receive the COVID-19 immunization. To protect the safety of expectant mothers and foetuses, government agencies and healthcare facilities should enhance the health monitoring of study participants.

5.6 COVID-19 Vaccines Improvement

Despite their ability to lower infection risk and patient mortality, COVID-19 vaccines now have issues with decreasing neutralizing activity of the variants and adverse events linked to immunization. The effectiveness of vaccines may be increased, and the incidence of adverse effects may be substantially decreased by adopting mix-and-match vaccinations and creating novel vaccines, such as virus-like particles and nanoparticle vaccines, enhancing current vaccine adjuvants, and altering the immunization route.

The original mRNA vaccines used in France offer immunocompetent persons 50 years of age and older high levels of prevention against both symptomatic SARS-CoV-2 infection and serious COVID-19 outcomes after two doses of the full primary vaccination. When mRNA is being endocytosed and moving across negatively charged cell membranes, nanoparticles can surround it and shield it from enzymatic destruction (Fig. 5.1). After escaping the nanoparticles, mRNA is liberated into the cytosol. This makes it possible for foreign proteins to appear on MHC II without having to break free from the endosome and enter the cytoplasm. The MHC I pathway is where pathogenic or intended-purpose endogenous proteins are first seen. The endoplasmic reticulum's MHC I molecules carry the smaller peptides that are produced when proteins are broken down in the proteasome. CD8+ T lymphocytes are presented with the peptide-MHC I complex at the cell surface in a sequential manner [40]. As of March 28, 2022, World Health Organization has authorized the emergency (urgent) utilization of eleven vaccinations, as well as three vaccines from India. These include inactivated vaccines, mRNA vaccines, viral vector vaccines, and protein subunit vaccines (https://extranet.who.int/pqweb/vaccines/vaccinescovid-19-vaccine-eul-issued). According to the mixed vaccination results obtained from ZF2001 and CoronaVac, the latter has superior immunogenicity while the former is considerably safer. Furthermore, compared to homologous vaccination, Scientists discovered that the combination of CoronaVac and Ad5-nCoV can produce more counteract antibodies and offer greater immunity (Zhu et al). Furthermore, Zhu et al. discovered that compared to homologous vaccination, the combination of CoronaVac and Ad5-nCoV can produce more neutralizing antibodies and provide superior protection. Traditional vaccines are not as effective at protecting the immune system as new vaccine platforms like mRNA vaccines. Nevertheless, these vaccinations are less effective at neutralizing variants, particularly the Beta and Delta. mRNA vaccines may not have the same neutralizing activity as nanoparticle vaccines, offering a new avenue for vaccine devel-

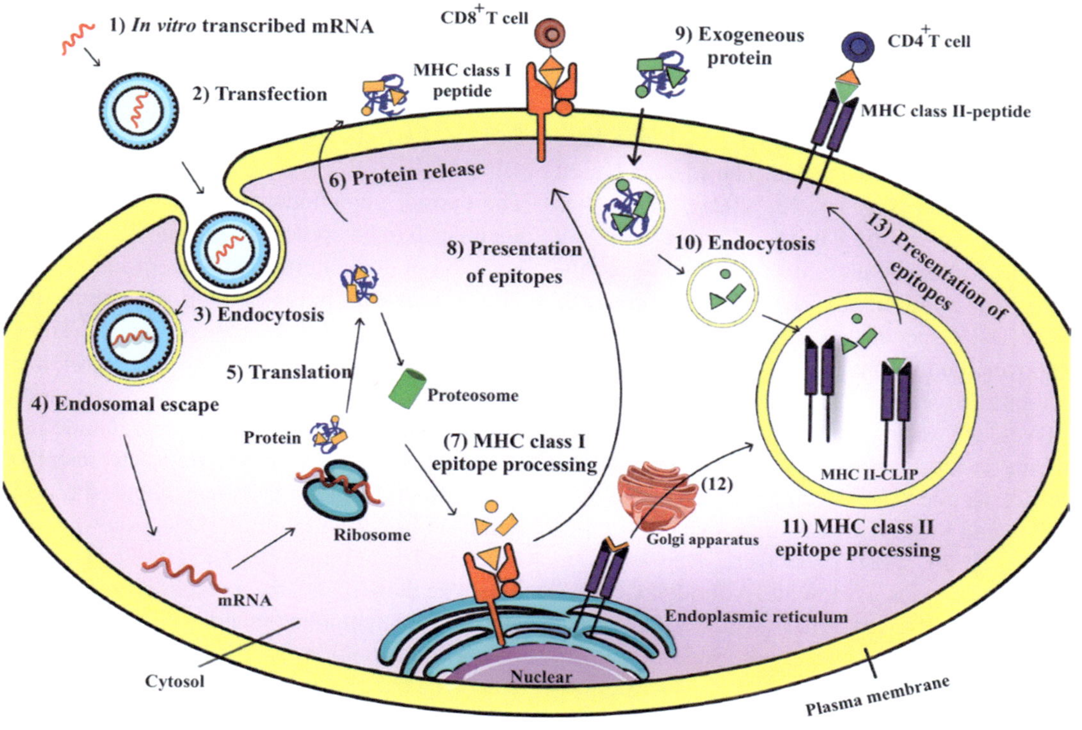

Fig. 5.1 mRNA vaccines employing positively charged lipid nanoparticles. (Permitted use from Elsevier with license number 5940670590858 with effect from Jan second, 2025)

opment. A ferritin skeleton and 24 polymer of SARS-CoV-2 RBD nanoparticles make up the nanoparticle vaccination developed by Ko et al. Cross-neutralizing antibody responses to SARS-CoV, bat coronavirus, and SARS-CoV-2, including Beta, Alpha, and Gamma forms, were brought on by the vaccination. The vaccine-induced DH1041-DH1045 strong neutralizing antibody exhibited neutralizing action against several mutations, including E484K, K417N, and N501Y. An immunogen made of 60 (SARS-CoV-2 S protein RBDs) was created by Walls et al. as a self-assembled protein nanoparticle. Even at modest dosages, the immunogen(antigen) can target many immunological epitopes, and yet produce large amounts of neutralizing antibody production. Furthermore, in contrast to conventional vaccines, nanoparticles have a long half-life in B-cell follicles, generating a persistent germinal centre reaction that guarantees robust antibody production.

To increase the effectiveness of the vaccine, additional significant viral proteins, such as S protein RBD, hemagglutinin, and others, could be added and operate as the physiologically appropriate trimeric viral spike form by self-assembly function of ferritin Consequently, nanoparticle vaccines are anticipated to be a significant component of the next COVID-19 vaccines due to their ability to optimize antigen packaging and generate a more robust and sustained immune response [39].

5.7 Vaccination Campaign Goals and Challenges

The goal of the global COVID-19 vaccine effort and struggle to vaccinate as many people as achievable is to prevent the disease, decrease the rate of infection, minimize terrible outcomes, and allow individuals to return to their regular lives. The objective of the continent's immunization campaign is to vaccinate more than 60% of the African population to achieve herd immunity, however most African nations started their national rollout of the COVID-19 vaccine later than other parts of the world. The African Union announced on January 13, 2021, that it has acquired 300 million doses of the COVID-19 vaccine for use by nations throughout the continent. In addition to obtaining millions of COVID-19 vaccines for the world's poor and impoverished countries, this huge amount of agreement was a stand-alone attempt of the COVID-19 vaccine Global Access (COVAX). Africa is the least vaccinated continent in the world, despite these vaccine procurement agreements and vaccine availability, even though as of August 24, 2021, there were over 7 million COVID-19 cases worldwide. Although COVAX is assisting in the purchase of COVID-19 vaccines for developing and underdeveloped nations, 60% of the funds spent on vaccinations go towards operational expenses like vaccine transportation, storage facilities, vaccinators' training, protective gear for vaccinators and other support personnel, and vaccination supplies. Many countries are considering requiring the COVID-19 vaccine passport for their citizens to enter public areas. Countries including the United Kingdom, France, Israel, China, and every nation in the European Union, for instance, have outlined plans to make COVID-19 vaccine passports a requirement for entry into a few designated public areas. Currently, nations like the United States and Australia have decided against requiring their residents to carry a COVID-19 vaccine passport, while some US states have COVID-19 vaccination mandates [41]. According to the World Health Organization (WHO), vaccine doubtfulness is a behaviour that is strike by a variety of factors, such as acceptance (do not see the need for a vaccine, do not value the vaccine), convenience (access), and confidence (do not trust the vaccine or provider). A diverse collection of people who are unsure about certain vaccines or vaccinations in general are known as vaccine-hesitant people. Vaccine-hesitant people may accept all vaccines but still be worried about them, reject, or postpone some immunizations but accept others, or reject all vaccines [42].

5.8 Conclusion

The current guidelines aimed at that infectious agent need to be updated to prevent the complete isolation of communities, groups, or individuals. This is necessary in the event of a pandemic and the ongoing spread of an infectious agent. This is done by increasing the capacity for control at an international level and gathering evidence on the virus's infectivity and elimination. Guidelines may then be revised considering new scientific and epidemiological evidence about the pandemic, considering the information known at the time of release. It may be first identified in secretions like those of the upper respiratory tract, possibly before symptoms manifest, and the highest risk of transmission is in the time immediately preceding the development of symptoms. Since it is now easier than ever to move across cities and countries, it is vital to control travellers beforehand. Thus, the following is a summary of the key elements of pandemic management: (a) implementation of epidemiological surveillance methods and procedures, including case definition, diagnostic confirmation, contact tracing, telemedicine, and psychological support; (b) management of certain subgroups with special needs, including pregnant women, children, patients with comorbidities, and others; (c) prevention and containment measures, including prevention measures for the general public, prevention measures for health workers, and personal protective equipment (PPE) for health workers; (d) cooperation between the various National Institutions and international public health strategies, including WHO, CDCs USA, ECDC, and others; and (e) the significance of accurate communication between the population, institutions, and health workers.

Acknowledgement The authors AR and GBJ are thankful to the National Research Foundation, South Korea for providing financial assistance.

Funding National Research Foundation, South Korea.

References

1. Lappeman J, Munyai K, Kagina BM. Negative sentiment towards COVID-19 vaccines: a comparative study of USA and UK social media posts before vaccination rollout. F1000Research. 2021;10(472):472.
2. Zolfaghari Emameh R, Nosrati H, Taheri RA. Combination of biodata mining and computational modelling in identification and characterization of ORF1ab polyprotein of SARS-CoV-2 isolated from oronasopharynx of an Iranian patient. Biol Proc (Online). 2020;22:8.
3. Pecetta S, et al. Immunology and technology of severe acute respiratory syndrome coronavirus 2 (SARS-CoV-2) vaccines. Pharmacol Rev. 2022;74(1):313–39.
4. Zolfaghari Emameh R, et al. Expansion of single cell transcriptomics data of SARS-CoV infection in human bronchial epithelial cells to COVID-19. Biol Proc (Online). 2020;22:16.
5. Wang X, Du Z, Johnson KE, Pasco RF, Fox SJ, Lachmann M, et al. The impacts of COVID-19 vaccine timing, number of doses, and risk prioritization on mortality in the US. medRxiv2021. 2021:2021.01.18.21250071.
6. Mathieu E, Ritchie H, Ortiz-Ospina E, Roser M, Hasell J, Appel C, et al. A global database of COVID-19 vaccinations. Nat Hum Behav. 2021;5(7):947–53.
7. Engelbrecht FA, Scholes RJ. Test for Covid-19 seasonality and the risk of second waves. One Health. 2021;12:100202.
8. Hung IFN, Poland GA. Single-dose Oxford-AstraZeneca COVID-19 vaccine followed by a 12-week booster. Lancet. 2021;397:854–5.
9. Moghadas SM, Vilches TN, Zhang K, Nourbakhsh S, Sah P, Fitzpatrick MC, Galvani AP. Evaluation of COVID-19 vaccination strategies with a delayed second dose. PLoS Biol. 2021;19:e3001211.
10. Spencer AJ, McKay PF, Belij-Rammerstorfer S, Ulaszewska M, Bissett CD, Hu K, Samnuan K, Blakney AK, Wright D, Sharpe HR, et al. Heterologous vaccination regimens with self-amplifying RNA and adenoviral COVID vaccines induce robust immune responses in mice. Nat Commun. 2021;12:2893.
11. Wolff J, Atuire C, Bhan A, Emanuel E, Faden R, Ghimire P, Greco D, Ho CWL, Kochhar S, Moon S, et al. Ethical and policy considerations for COVID-19 vaccination modalities: delayed second dose, fractional dose, mixed vaccines. BMJ Glob Health. 2021;6:e005912.
12. Bradley BT, Maioli H, Johnston R, Chaudhry I, Fink SL, Xu H, Najafian B, Deutsch G, Lacy JM, Williams T, Yarid N, Marshall DA. Histopathology, and ultrastructural findings of fatal COVID-19 infections in Washington State: a case series. Lancet. 2020;396:320–32. https://doi.org/10.1016/S0140-6736(20)31305-2.
13. Unudurthi SD, Luthra P, Bose RJC, McCarthy J, Kontaridis MI. Cardiac inflammation in COVID-19: lessons from heart failure. Life Sci. 2020;260:118482. https://doi.org/10.1016/j.lfs.2020.118482.
14. Wichmann D, Sperhake JP, Lütgehetmann M, Steurer S, Edler C, Heinemann A, Heinrich F, Mushumba H, Kniep I, Schroder AS, Burdelski C, de Heer G, Nierhaus A, Frings D, Pfefferle S, Becker H, Bredereke-Wiedling H, de Weerth A, Paschen HR, Sheikhzadeh-Eggers S, Stang A, Schmiedel S, Bokemeyer C, Addo MM, Aepfelbacher M, Püschel K, Kluge S. Autopsy findings and venous thromboembolism in patients with COVID-19: a prospective cohort study. Ann Intern Med. 2020;173:268–77. https://doi.org/10.7326/M20-2003.
15. Chen T, Wu D, Chen H, Yan W, Yang D, Chen G, Ma K, Xu D, Yu H, Wang H, Wang T, Guo W, Chen J, Ding C, Zhang X, Huang J, Han M, Li S, Luo X, Zhao J, Ning Q. Clinical characteristics of 113 deceased patients with coronavirus disease 2019: retrospective study. BMJ. 2020;368:1091.
16. Sadhukhan P, Ugurlu MT, Hoque MO. Effect of COVID-19 on lungs: focusing on prospective malignant phenotypes. Cancer. 2020;12:3822.
17. Chau T-N, Lee K-C, Yao H, Tsang T-Y, Chow T-C, Yeung Y-C, Choi K-W, Tso Y-K, Lau T, Lai S-T, Lai C-L. SARS-associated viral hepatitis caused by a novel coronavirus: report of three cases. Hepatology. 2004;39:302–10.
18. Zhou S, Wang Y, Zhu T, Xia L. CT features of coronavirus disease 2019 (COVID- 19) pneumonia in 62 patients in Wuhan, China. Am J Roentgenol. 2020;214:1287–94.
19. Hussain A, Via G, Melniker L, Goffi A, Tavazzi G, Neri L, Villen T, Hoppmann R, Mojoli F, Noble V, Zieleskiewicz L, Blanco P, Ma IWY, Wahab MA, Alsaawi A, Al Salamah M, Balik M, Barca D, Bendjelid K, Bouhemad B, Bravo-Figueroa P, Breitkreutz R, Calderon J, Connolly J, Copetti R, Corradi F, Dean AJ, Denault A, Govil D, Graci C, Ha Y-R, Hurtado L, Kameda T, Lanspa M, Laursen CB, Lee F, Liu R, Meineri M, Montorfano M, Nazerian P, Nelson BP, Neskovic AN, Nogue R, Osman A, Pazeli J, Pereira-Junior E, Petrovic T, Pivetta E, Poelaert J, Price S, Prosen G, Rodriguez S, Rola P, Royse C, Chen YT, Wells M, Wong A, Xiaoting W, Zhen W, Arabi Y. Multi-organ point-of-care ultrasound for COVID-19 (PoCUS4COVID): international expert consensus. Crit Care. 2020;24:702.
20. Topol EJ. COVID-19 can affect the heart. Science. 2020;370:408–9.
21. Baksh M, Ravat V, Zaidi A, Patel RS. A systematic review of cases of acute respiratory distress syndrome in the coronavirus disease 2019 pandemic. Cureus. 2020;12:8188.
22. Yang X, Yu Y, Xu J, Shu H, Xia J, Liu H, Wu Y, Zhang L, Yu Z, Fang M, Yu T, Wang Y, Pan S, Zou X, Yuan S, Shang Y. Clinical course, and outcomes of critically ill patients with SARS-CoV-2 pneumonia in Wuhan,

China: a single-centered, retrospective, observational study. Lancet Respir Med. 2020;8:475–81.
23. Guan W, Liang W, Zhao Y, Liang H, Chen Z, Li Y, Liu X, Chen R, Tang C, Wang T, Ou C, Li L, Chen P, Sang L, Wang W, Li J, Li C, Ou L, Cheng B, Xiong S, Ni Z, Xiang J, Hu Y, Liu L, Shan H, Lei C, Peng Y, Wei L, Liu Y, Hu Y, Peng P, Wang J, Liu J, Chen Z, Li G, Zheng Z, Qiu S, Luo J, Ye C, Zhu S, Cheng L, Ye F, Li S, Zheng J, Zhang N, Zhong N, He J. Comorbidity, and its impact on 1590 patients with COVID-19 in China: a nationwide analysis. Eur Respir J. 2020;55:2000547.
24. Richardson S, Hirsch JS, Narasimhan M, Crawford JM, McGinn T, Davidson KW, Barnaby DP, Becker LB, Chelico JD, Cohen SL, Cookingham J, Coppa K, Diefenbach MA, Dominello AJ, Duer-Hefele J, Falzon L, Gitlin J, Hajizadeh N, Harvin TG, Hirschwerk DA, Kim EJ, Kozel ZM, Marrast LM, Mogavero JN, Osorio GA, Qiu M, Zanos TP. Presenting characteristics, comorbidities, and outcomes among 5700 patients hospitalized with COVID-19 in the New York city area. JAMA J Am Med Assoc. 2020;323:2052–9.
25. Shi S, Qin M, Shen B, Cai Y, Liu T, Yang F, Gong W, Liu X, Liang J, Zhao Q, Huang H, Yang B, Huang C. Association of cardiac injury with mortality in hospitalized patients with COVID-19 in Wuhan, China. JAMA Cardiol. 2020;5:802–10.
26. Yang J, et al. Prevalence of comorbidities and its effects in coronavirus disease 2019 patients: a systematic review and meta-analysis. Int J Infect Dis. 2020;94:91–5.
27. Turner AJ, Hiscox JA, Hooper NM. ACE2: from vasopeptidase to SARS virus receptor. Trends Pharmacol Sci. 2004;25:291–4.
28. Zheng YY, Ma YT, Zhang JY, Xie X. COVID-19 and the cardiovascular system. Nat Rev Cardiol. 2020;17:259–60.
29. Agarwal A, Chen A, Ravindran N, C. To, Thuluvath PJ. Gastrointestinal and liver manifestations of COVID-19. J Clin Exp Hepatol. 2020;10:263–5.
30. Hundt MA, Deng Y, Ciarleglio MM, Nathanson MH, Lim JK. Abnormal liver tests in COVID-19: a retrospective observational cohort study of 1,827 patients in a major U.S. hospital network. Hepatology. 2020;72:1169–76.
31. Cai Q, Huang D, Yu H, Zhu Z, Xia Z, Su Y, Li Z, Zhou G, Gou J, Qu J, Sun Y, Liu Y, He Q, Chen J, Liu L, Xu L. COVID-19: abnormal liver function tests. J Hepatol. 2020;73:566–74.
32. Zhang C, Shi L, Wang F-S. Liver injury in COVID-19: management and challenges. Lancet Gastroenterol Hepatol. 2020;5(5):428–30.
33. Cheung KS, Hung IFN, Chan PPY, Lung KC, Tso E, Liu R, Ng YY, Chu MY, Chung TWH, Tam AR, Yip CCY, Leung KH, Fung AYF, Zhang RR, Lin Y, Cheng HM, Zhang AJX, To KKW, Chan KH, Yuen KY, Leung WK. Gastrointestinal manifestations of SARS-CoV-2 infection and virus load in fecal samples from a Hong Kong cohort: systematic review and meta-analysis. Gastroenterology. 2020;159:81–95.
34. Su S, Du L, Jiang S. Learning from the past: development of safe and effective COVID-19 vaccines. Nat Rev Microbiol. 2021;19:211–9.
35. Estes JD, Wong SW, Brenchley JM. Nonhuman primate models of human viral infections. Nat Rev Immunol. 2018;18:390–404.
36. Muñoz-Fontela C, et al. Animal models for COVID-19. Nature. 2020;586:509–15.
37. Li T, et al. Mask or no mask for COVID-19: a public health and market study. PLoS One. 2020;15(8):e0237691.
38. Wang Y, Deng Z, Shi D. How effective is a mask in preventing COVID-19 infection? Med Dev Sens. 2021;4(1):e10163.
39. Li M, Wang H, Tian L, Pang Z, Yang Q, Huang T, Fan J, Song L, Tong Y, Fan H. COVID-19 vaccine development: milestones, lessons, and prospects. Signal Transduct Target Ther. 2022;7(1):146.
40. Mirtaleb MS, Falak R, Heshmatnia J, Bakhshandeh B, Taheri RA, Soleimanjahi H, Emameh RZ. An insight overview on COVID-19 mRNA vaccines: advantageous, pharmacology, mechanism of action, and prospective considerations. Int Immunopharmacol. 2023;117:109934.
41. Ayenigbara IO, Adegboro JS, Ayenigbara GO, Adeleke OR, Olofintuyi OO. The challenges to a successful COVID-19 vaccination programme in Africa. Germs. 2021;11(3):427–40. https://doi.org/10.18683/germs.2021.1280.
42. Troiano G, Nardi A. Vaccine hesitancy in the era of COVID-19. Public Health. 2021;194:245–51. https://doi.org/10.1016/j.puhe.2021.02.025.
43. Nwagwu CS, Ugwu CN, Ogbonna JDN, Onugwu AL, Agbo CP, Echezona AC, et al. Recent and advanced nano-technological strategies for COVID-19 vaccine development. In: Methods in microbiology, vol. 50. Academic; 2022. p. 151–88.

6 Economic Crisis in COVID-19 and Strategies for Recovery

Qazi Noorul Mateen, Qazi Ghazala Mateen, Ankita Raghav, Virendra Singh, Supriya Raghav, Santosh Kumar, and Poonam Khanna

6.1 Introduction

In December 2019, there were many cases reported of pneumonia of an unknown cause in Wuhan, China. Within 2 weeks, this respiratory virus was identified as a novel coronavirus related to SARS-CoV, named severe acute respiratory syndrome coronavirus 2 (SARS-CoV-2) [17]. This virus spread rapidly to all countries, causing a severe toll on the lives and livelihoods of the population. The COVID-19 pandemic is termed as the worst global crisis after World War 2 [33]. The World Health Organization (WHO) declared it a Public Health Emergency of International Concern (PHEIC) on January 30, 2020.

By August 2020, there were 30 million confirmed cases and over 850,000 deaths due to COVID worldwide. This continued to escalate to 184 million confirmed cases and over 4 million deaths by the end of the first half of 2021 [40]. The COVID-19 pandemic emerged as an unprecedented global crisis, profoundly affecting the public healthcare systems across nations, economies, and societies. The crisis was distinct in its global scale, speed of impact, and multifaceted effects, setting it apart from earlier crises such as SARS (2003) and MERS (2012), which remained largely regional [2].

From February 21 to March 24, 2020, the COVID-19 pandemic spread around the world, and the number of cases was continuously rising [58]. As the cases were increasing worldwide, the news around COVID-19 cases took over any other news. This continued broadcasting of pandemic-related news created a sense of anxiety and panic among people and the government [7]. The governments of different countries took strict measures such as lockdowns, restricted movements, and halting of other activities [12]. This further created the insecurity among people and influenced their investment decisions, resulting in high volatility in stock markets and trading activities [67]. There were immediate economic repercussions when lockdowns and social distancing measures were put into place. One impor-

Authors "Qazi Noorul Mateen", "Qazi Ghazala Mateen", and "Ankita Raghav" have equally contributed to this chapter.

Q. N. Mateen
Liverpool John Moores University, Liverpool, UK

Q. G. Mateen
Department of Economics, Aligarh Muslim University, Aligarh, India

A. Raghav (✉) · V. Singh
Department of Economics, IIMT University, Meerut, India

S. Raghav · S. Kumar
School of Public Health, Poornima University, Jaipur, India

P. Khanna
School of Public Health, Post Graduate Institute of Medical Education and Research, Chandigarh, India

A. Raghav et al. (eds.), *COVID-19 Impact on Economy, Environment and Healthcare*,
https://doi.org/10.1007/978-981-95-1196-9_6

tant industry for many economies, tourism, was especially impacted [1]. Following initial reports from China, few anticipated the scale of economic disruption that would follow [12]. And China being the major center for global manufacturing affected its own productivity, which in turn affected other interconnected economies around the world [63].

In the European Union (EU), the first reported case was on January 24, 2020 and the peak for the first wave came in April 2020 [59]. Following this wave, the infection rate rose once more in the fall of 2020 and stayed constant until the beginning of 2021. In the EU, the GDP rate decreased by 6.1% as a result of the measures taken to stop the virus [34]. India reported its first case on January 27, 2020, and its first COVID-19-related death on March 12, 2020. India took the most severe action of any nation, enforcing a total lockdown on March 25. As a result, fewer cases were reported by April 15, 2020, than anticipated. By April 15, 2020, there were only 11,738 cases, compared to 820,000 if the lockdown had not been implemented [16]. This worked to contain the infection in the early stages, but then the economy started to deteriorate, and India's GDP fell by a staggering 24.4% between April and June 2020 [56]. The United States reported its first case on January 19, 2020, in Washington, and due to a lack of preparedness, there is an unexpected rise in the number of cases [31, 65]. The US Department of Labor reported the unemployment rate jumped to 14.7% in April 2020 [65].

The major economic recessions from the past were mainly the result of either a demand-side or a supply-side factor. For instance, the Great Depression of the 1930s was due to supply-side disruptions; the 1970s oil shock too was due to cost-push inflation. The 2008 financial crisis was the result of demand-driven disruptions. The COVID-19 pandemic, on the other hand, disrupted supply chains while simultaneously stifling consumer demand as production halted and consumption patterns shifted drastically [10].

The COVID-19 pandemic has far-reaching social, economic, and mental health consequences on individuals, societies, and economies across the globe. It revealed the inequalities within different countries, and it further widened this gap between advanced and emerging economies. It also revealed the unpreparedness of households, businesses, and governments for such an uncertain health emergency and economic shock, as the studies based on pre-crisis data suggested that average businesses could cover less than 55 days on cash revenue, and more than 50% of households cannot complete their basic consumption demands for more than 3 months in an event of an economic crisis. For workers with only a primary education, temporary unemployment was higher in 70% of all countries in 2020, according to survey data. Youth, women, independent contractors, and casual workers with less formal education also experienced greater income losses. Due to their greater likelihood of working in fields more impacted by lockdown regulations and social distancing, women were specifically affected by income and job losses [61].

The response to this crisis was also large and impulsive in terms of policies and healthcare spending, which dramatically increased public and private debt on the global economy. In emerging economies, a handsome number of households and firms were already burdened with significant debt even prior to the COVID-19 crisis, and this made the situation worse for them in their ability to generate revenue and repay debt [30, 61].

The trends among businesses are similar. The pandemic caused higher income losses for smaller businesses, informal enterprises, and businesses with less access to formal credit. Microenterprises were able to cover costs for 50 days, medium-sized businesses for 59 days, and larger companies for 65 days. Additionally, micro, small, and medium-sized businesses are over-represented in the sectors utmost affected by the crisis. These generally include retail, personal services, and lodging and food services [61].

6.2 Major Economic Impacts

In this section, we examine the profound and far-reaching economic consequences of the COVID-19 pandemic across various sectors of the global economy. One of the fastest and most widespread stock market crashes in history was the first disruption brought on by COVID-19, with big indices like the FTSE, Dow Jones, and Nikkei seeing precipitous drops in early 2020. While some industries, such as food, chemicals, and natural gas, demonstrated resiliency and even made progress, others experienced significant setbacks.

The study looks at how China's dominant position in manufacturing had a major impact on global supply chains, causing major production delays and shortages in a variety of industries. During the ensuing employment crisis, the United States alone lost over 26 million jobs in just 5 weeks starting in mid-March 2020. Significant revenue losses forced major hotel chains to restructure their operations in the tourism sector, and a paradigm shift toward digital platforms occurred in the entertainment sector, especially in the film industry.

The strain on international healthcare systems is also examined in this section, with a focus on infrastructure problems and workforce impacts. Each subsection offers comprehensive data, case studies, and analysis to show how the pandemic caused interrelated problems in every area of the global economy, paying special attention to how it disproportionately affected workers in the informal sector, such as women, young people, and migrant workers.

Through a thorough analysis of these crucial areas, this section offers a comprehensive understanding of how COVID-19 changed the local and global economic landscape.

6.2.1 Stock Market Crash and Volatility

As COVID-19 made headlines in February 2020, its impact on the stock market was also noticeable. Since the circuit breaker's inception in 1987, it has only been triggered once, in 1997, but in March 2020, the US stock market set it off four times in 10 days. Global stock markets saw a coordinated shock in February–March 2020 that was faster and more extensive than even the 2008 financial crisis, making the initial stock market impact of COVID-19 one of the fastest and most pervasive financial disruptions in history [41]. The stock markets in Europe and Asia have also plummeted in tandem with the US stock crash [68]. On March 12, 2020, the UK's main index experienced its worst day since 1987, falling more than 10%. Japan's stock market fell more than 20% from its peak in December 2019 [68]. The FTSE, Dow Jones, and Nikkei all experienced sharp drops, and the S&P 500 fell by a third in a few short weeks [64].

A study by Mazur et al. [44] analyzed data from S&P firms during March 2020 using stock prices and trading volume. Due to panic selling of stocks, the unexpected winners and losers were exposed. The natural gas companies experienced gains of 11–17% due to reduced oil production, and the firms related to food and chemicals also saw jumps of 20% in 1 day. On the other hand, stock prices crashed for companies dealing in the hospitality, real estate, and entertainment sectors. These losses were observed to be even higher than 60%. Businesses in the food distribution, software, and healthcare sectors did well, generating returns of over 20% per month. This is particularly due to the surge in demand for remote solutions for work and basic food supply needs.

Another study by Baek et al. [6] analyzed the US stock market from January 2, 2020, to April 30, 2020, using a two-regime Markov switching model and found out that the volatility is more perceptive to the news of COVID-19 than the economic indicators. There is a negative bias observed in the stock market trend as the impact of COVID-19 negative news such as death numbers was twice as efficacious negatively than positive COVID-19 news such as recoveries.

Using data from the United States, Japan, India, Germany, China, the United Kingdom,

France, Italy, Brazil, and Canada from January 1, 2019, to June 30, 2020 [15] found that the COVID-19 period was linked to negative mean returns for all market indices and higher volatility on a global scale.

The 225 major companies listed on the Tokyo Stock Exchange are represented by the Nikkei, 30 well-known American companies across a range of industries are tracked by the Dow Jones, and 100 major companies listed on the London Stock Exchange are represented by the FTSE 100. As shown in Fig. 6.1, all three indexes—Nikkei (Tokyo), FTSE (London), and Dow Jones (New York)—show a sharp decline by late February 2020 and March 2020. The indexes were trading at around 0% at the beginning of the year, but as the COVID-19 news began to make headlines, stock prices fell sharply, causing a global slowdown and investor and retailer panic.

Despite higher volatility and lower overall stock returns, the impact of economic challenges varied across nations, reflecting differences in their economic resilience, policy responses, and economic size. In South Korea, for example, the withdrawal of foreign capital led to currency depreciation and increased market instability, while research employing cross-sectional deviation models revealed that investors in Europe exhibited herding behavior [25, 32].

6.2.2 Supply Chain Disruptions

Global supply chains have made countries more vulnerable to risks that can go beyond territorial boundaries, and the world was unprepared for a crisis of the magnitude of COVID-19 [66]. Since China is the world's largest exporter and the focal point of global manufacturing, the COVID-19 pandemic has had a knock-on effect on all of the interconnected economies in the world. In the first month of COVID-19, which was February 2020, the sectors with the highest exposure to Chinese imports did not differ significantly from the less exposed sectors, according to a study by Meier and Pinto [45]. However, in March, when COVID-19 began to peak in China and there was a shortage of workers at work, warehouses, and loading and unloading facilities, a significant difference between the high-exposed sectors and the less-exposed sectors can be seen in Fig. 6.2. In comparison to January 2020, the production decline in industries with greater exposure to China was 2% larger by March and increased to 16% in April [45].

According to the study by Pujawan and Bah [50], after realizing how the supply chain can be disrupted in any crisis, such as COVID-19, the following areas should be the focus of attention. The emphasis should be on safety, the supply chain should be increasingly digitized, the supply chain should gradually become more localized,

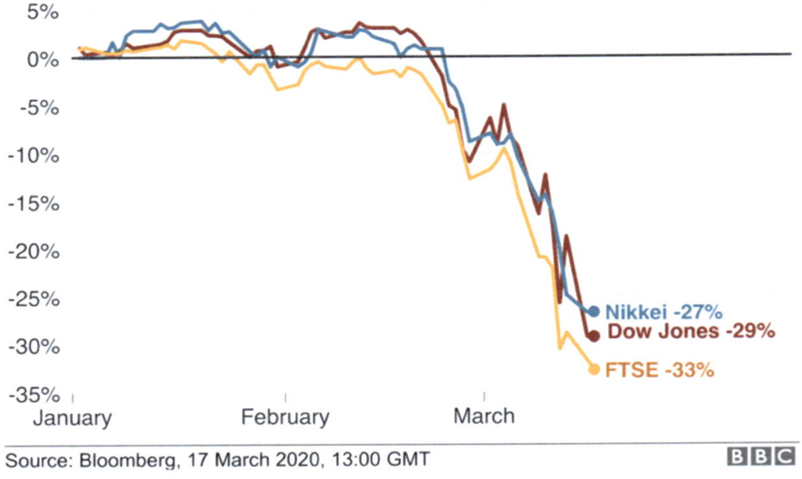

Fig. 6.1 COVID-19 impact on different stock markets since the start of the outbreak

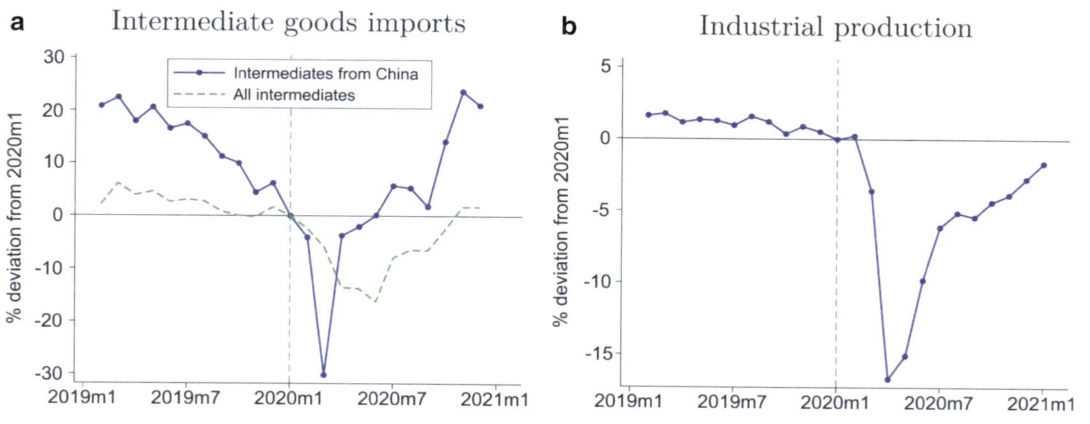

Fig. 6.2 Impact of COVID-19 on global supply chain and industrial production (2019 m1–2021 m1). (**a**) Comparison of intermediate goods imports from China vs all intermediates. (**b**) Changes in industrial production. (Source: COVID-19 Supply Chain Disruptions [45] (used terms and conditions of Common Creative License 4.0))

and efficiency in vaccine production and distribution should be rethought. Although the recommendations for boosting supply chain flexibility and resources prior to COVID-19 are still relevant today, there is currently a greater push for local sourcing and digital tools.

Additionally, COVID-19 has affected the global food supply chain, putting a financial strain on the government to feed the hungry. Although there was no direct impact on food production like there was for other diseases like bird flu or E. coli, lockdown and COVID restrictions forced farmers to destroy or discard their food supplies. Several dairy farmers dumped liters of milk, and food products have a short shelf life, so they must get to the customer by a certain date [9]. The supply was disrupted, but the demand for food, specifically bread, increased by 76% and vegetables by 52% within a week of declaring the COVID-19 pandemic [18].

The global auto market was impacted by the supply chain disruption because two-thirds of the auto parts were produced in 11 Chinese provinces. The Hubei province supplies for the US, Europe, and South Korean markets, and this disruption caused the loss of 350,000 units of production at the start of the crisis, i.e., February 2020, and this number reached 1.7 million by March 2020 [48]. Hyundai Motor Company was forced to close the domestic factory and similarly, US and European automakers have faced significant challenges in their production and business [48].

The study by Moosavi et al. [46] has suggested that areas should be points of focus after realizing how the supply chain can be disrupted in any crisis like COVID-19. There should be a focus on safety, increasing digitization of the supply chain, gradually shifting toward a localized supply chain, and rethinking efficiency and production and distribution of vaccines. There were already recommendations in place before COVID-19 for building more resources and flexibility in the supply chain, and these recommendations still hold true. Now the push for local sourcing and digital tools is more.

6.2.3 Employment Crisis

The pandemic caused a never-before-seen crisis in employment, with varying effects in different industries and geographical areas [62]. The manufacturing and services sectors were hardest hit, with total closures documented in sectors like tourism and hospitality that had previously withstood economic downturns [48]. As an initial impact profit margins and sales declined, many people started losing employment and money [24]. This impacted the people both economically and mentally. Although lockdowns have far-reaching economic effects, some groups of people are disproportionately affected, particularly those in the informal sector of the Indian economy. This group includes women, young people, migrant workers, small traders, independent contractors, and daily wage workers [3].

The United States has also seen one of the largest layoffs, with more than 26 million jobs being lost in just 5 weeks from mid-March 2020 to April 2020. The unemployment rate of the country also reached 4.4% in the month of March (as shown in Fig. 6.3a). In the United States, the unemployment rate was worse for females initially; it came to the same level after 1 year. The industry-wise unemployment data as shown in Fig. 6.3b shows that the leisure and hospitality industry faces the highest unemployment with a peak of 39.3% in April 2020. The reason for this is quite obvious, as the travel restrictions were imposed and social distancing measures were observed. By February 2021, this had somewhat improved, but the unemployment rate remained at 13.5%, which is significantly worse than any other sector during this period. Another industry where the unemployment got worse in the year is mining and extraction. The unemployment rate increased from 5.4% in February 2020 to 19.3% in February 2021 [19]. The industries associated with financial activities, government, and education and health services showed lesser unemployment. Those with lower wages and within leisure, hospitality, and service jobs were more likely to lose their jobs. The unemployment improved year over year in most of the sectors except agriculture, which has not seen the significant increase or decrease in unemployment. 158.8 million individuals were employed in the United States on average in the fourth quarter of 2022 [26].

Though the European Union (EU) countries have the issue of youth unemployment, the EU aims to lessen it in order to boost their economy, but in 2020, the COVID-19 pandemic prompted stringent measures to contain the virus, resulting in a sharp increase in unemployment. In nations like Greece, Spain, and Italy, youth unemployment was already high, and COVID-19 worsened the situation, but also in countries like the Netherlands, Poland, and the Czech Republic, where unemployment was relatively less pre-COVID-19, there was a sharp increase in unemployment post-COVID-19. The youth unemployment rate in the Czech Republic increased to 2.19 times by the end of 2020 with the end of 2019. Similarly,

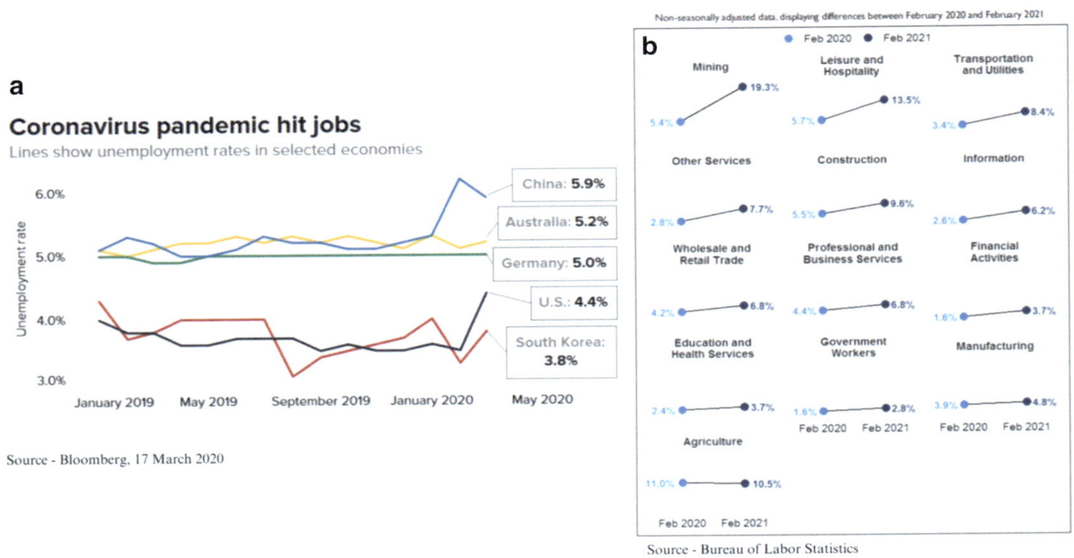

Fig. 6.3 (**a**) Monthly unemployment rate during COVID-19 (Jan 2019–May 2020); (**b**) industry-wise unemployment rates in the United States (February 2020–February 2021)

in Estonia, it increased to 2.5 times from the 2019 level [39].

During the peak COVID-19 years, the unemployment rate in India increased drastically. Lockdowns, business shutdowns, economic burdens on firms, and mass firings of employees made millions of people jobless. There were also cases of underemployment in different sectors of the economy. Figure 6.4, which illustrates the severity of the unemployment rate, shows that the overall unemployment rate peaked between March and April 2020, rising from 8.75 to 23.52 respectively. The lockdown was somewhat loosened starting in June 2020, which helped the unemployment rate, which had previously been above 20%, drop to about 10% in June and then further improve to 7.4% a month later. For the following few months, this stayed at about 7% until the COVID-19 second way strike in April 2021, which led to the imposition of another lockdown in numerous states. The unemployment rate rose once more as a result, reaching roughly 12% overall. Given that the impact of COVID-19 was worst in India's second wave, the impact on urban unemployment was worse, reaching about 15%, the second-highest level since the first lockdown announcement. In July 2021, unemployment returned to 8.3% as the situation improved and the second wave came to an end [43].

In India, around 85–90% of the population works in the informal sector, where there are no provisions such as job security and a basic wage rate [37]. The sudden, stringent lockdown by the government left millions of migrant workers unemployed, with no money or food to survive. Their situation got so bad that they had no option but to travel back to their villages under travel restrictions, which even cost many of them their lives. The major developed cities like Maharashtra, Gujarat, Karnataka, and Tamil Nadu have higher positive cases of COVID reported, and thus it affected the free mobility of labor flow from states with high surplus amounts of labor like Bihar, Jharkhand, and West Bengal. Their migrant labor lost their livelihoods due to industries and factory closures, and they migrated to their native places [52, 54].

This crisis also had a hard hit on the women migrants' loss of livelihood and income by resulting in severe debt, burden of responsibility, and insufficient financial and social support [55].

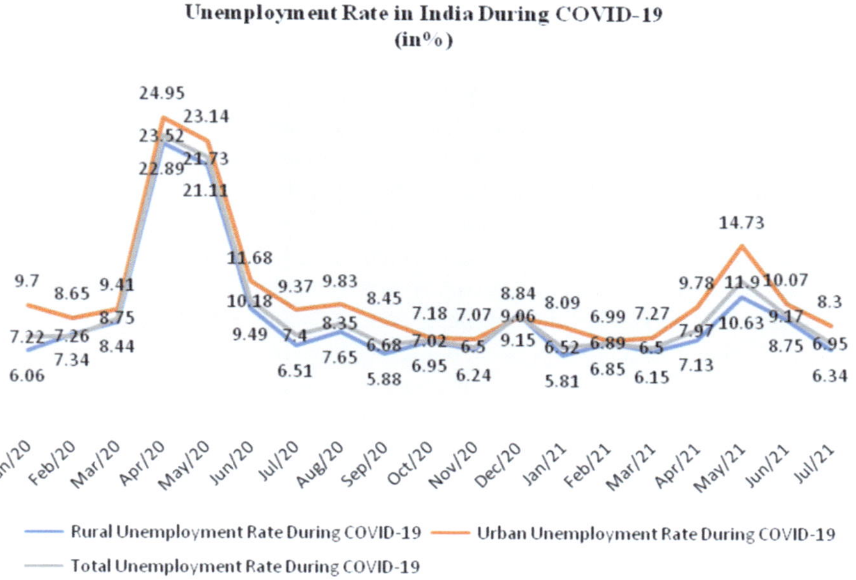

Fig. 6.4 Unemployment rate in India during COVID-19 (Jan 2020–July 2021). (Source: Adapted from Magesh Kumar et al. [42, 43])

The lockdown employment pattern made the women's labor force more vulnerable and jobless than men, and many of them never returned to work [21].

The mental health is also damaged by job loss, illness, and despair [22]. The behavior of the population was also affected due to anxiety, stress, and exhaustion [14].

6.2.4 Economic Loss in the Tourism Industry

The tourism industry across the world suffered significant losses in revenue, and the effect was initially evident in China. China was a popular tourist destination and a sizable source market for travelers. Around US $73 billion was generated in revenue by the Chinese tourists who visited abroad for Lunar New Year 2019 [48]. Chinese tourists who visit Vietnam and Thailand were also significantly reduced, causing a loss of US $7.7 billion and US $3.1 billion, respectively, in the first quarter of 2020 [48].

In the tourism sector, hotel business suffered one of the highest losses. Lockdowns, social isolation, and stringent travel restrictions or prohibitions restrict mobility, which hurts hotel businesses overall and ultimately leads to job losses in the hospitality industry [29]. For instance, Marriott International hotel chain that has around 174,000 employees was forced to release tens of thousands of employees to bear their business losses [8]. The hotel chains realized the impact of the crisis much before it took its worst shape and they started to prepare themselves to keep the company moving and functioning. As a precaution, Hilton Worldwide took out a $1.75 billion revolving loan on March 5, 2020, to cover this economic shock [8]. This move showed the volatility in the market around that time and people were getting prepared to face the worst, which was yet to come.

The income per room was significantly declined even at the initial phase of COVID-19. There was 11.6% decline in US hotel industry income per room and this was even worse in Germany, Rome, and London. In Rome, the occupancy rate was 6% and in London it was 47%. For Germany the occupancy reduced by 36% [48]. As Hitlon took the revolving loan, likewise, other hotel corporations also demanded US $150 billion in direct employee help due to their loss of US $1.5 billion since mid-February [11]. For China the occupancy rate reduced by 89% from January 2020 [48].

6.2.5 Decline in Cinema

The entertainment industry made a paradigm shift due to lockdowns. Cinema and theater, one of the most economically significant and popularized sectors, had burnt a huge burden of social distancing norms. Film making and production halted. Many film releases were delayed in the pandemic years. Several of them were later released only on OTT platforms like Netflix or Amazon Prime. In 2020, the entertainment industry lost $5 million. Apart from this, due to shutdowns of theaters, around 120,000 below-the-line entertainment jobs were lost [48].

The pandemic had compelled the production houses to be more innovative and to adapt digital transformation, safety protocols adding an extra layer of cost. The traditional screening and popcorn culture got shifted to the screening and distribution of films on digital platforms [51]. In Italy, the disease spreading costs millions of euros every week. Musical concerts, live shows, exhibitions, and art shows all lose revenue [48].

One major speculation in the entertainment industry was whether it would receive stimulus packages from the governments or not due to the fact that some consider the entertainment sector contributes more to the financial and industrial sectors and less to the economy [60]. The International Alliance of Theatrical Stage Employees (IATSE) requested that the governments include the entertainment industry in the stimulus package. After the reopening of the economies in 2021, the entertainment industry resumed and performed well, but it didn't reach the pre-pandemic levels in terms of revenue and production [48].

6.2.6 Healthcare System Strain

The cost of COVID illness and hospital admissions directly impacts people's finances in India, where insurance is not widely used and nearly 60% of the population uses private hospitals. There were 43 million cases reported by the end of June 2022 that were directly related to treatment costs. In addition, there are indirect costs, such as the patient and caregivers' lost productivity [28].

Significant flaws in healthcare infrastructures around the world were made clear by the COVID-19 pandemic, which demonstrated how ill-equipped systems were to handle patient surges and continue providing standard medical services. Critical shortages were caused by supply chain interruptions, which especially affected the availability of drugs. For example, fentanyl shortages necessitated the use of morphine sulfate and hydromorphone as substitute analgesics, while general anesthetic shortages of propofol and midazolam necessitated the use of ketamine and lorazepam. Communities with lower economic status were especially affected by these disruptions, making access to medications and medical services even more difficult [27].

Medical professionals were under a lot of stress, as they had to handle situations that were more akin to battlefield medicine than normal practice. 53% of US doctors report feeling burned out, and 23% exhibit symptoms of depression, according to recent polls. Compared to primary care physicians in every other country polled, 43% of US primary care physicians reported lower practice revenue during the pandemic, indicating that the financial impact was especially severe for small medical offices. Though this trend may result in greater healthcare costs because larger medical groups tend to demand higher charges and "up code" patient severity more frequently, this financial pressure is anticipated to accelerate the trend of physicians leaving small practices and joining larger groups [20].

6.3 Strategies for Recovery

The recovery from the COVID-19 crisis required policy responses from individual countries and international cooperation. Different economies around the world experienced varying degrees of impact, had different capacities, and implemented diverse preventive measures. A one-size-fits-all approach was not feasible, leading to varied recovery strategies. These ranged from fiscal stimulus packages to innovative monetary policies, supported by international coordination. This section examines how nations and global institutions undertook their recovery efforts, focusing on both individual responses and collaborative initiatives by international organizations that contributed to their journey back to stability.

6.3.1 Policy Response and Comprehensive Planning

The COVID-19 pandemic hit all economies differently. The nature of the recession faced by countries was dependent upon their pre-existing conditions and vulnerabilities, their supply and demand patterns, the kind of preventive measures they adopted, etc. The road to recovery and global response to the pandemic is also different among economies, and no single specific policy for the crisis could be implemented worldwide [35].

In order to mitigate the immediate and short-term impact of the pandemic, various supervisory authorities, governments, and central banks introduced policy measures, including both fiscal and monetary policies. For the immediate relief, large fiscal stimulus packages were sanctioned to economies. Both budgetary fiscal measures, such as spending on healthcare, tax cuts, providing subsidies, and direct cash transfer programs to households and firms and non-budgetary fiscal measures such as loan by government and credit guarantees, etc., were implemented [4]. Advanced economies were the first ones to announce these fiscal measures and emerging economies followed their path gradually. At the same time central banks took measures that helped in easy flow of credit and to provide liquidity in the economy.

6.3.1.1 United States Policy Response

In the US economy, 916,000 jobs were added in March 2021, which helped in the recovery from the job loss due to the sudden onset of COVID-19 a year before. This reduced the job loss to 6% from the peak of more than 15% in April–May 2021. The unemployment continued to decline in 2022 also. The number of unemployed people in 2022 was 5.9 million, which was 900,000 less than the end of 2021. And the unemployment rate reached 3.6%, which is at par with the pre-COVID-19 level. The number of employed people in the fourth quarter averaged 158.8 million [26].

The United States of America spent trillions of dollars in relief packages to boost the pent-up demand. Some of the major relief packages included the CARES Act announced in March 2020, which provided $2.3 trillion supporting tax rebates, unemployment benefits, healthcare relief packages, and international aid. In the same year, in December, the Consolidated Appropriation Act was announced that provided an additional stimulus package of $868 for unemployment benefits and $600 stimulus checks, vaccines, and for education. In March 2021 US announced an American Rescue Plan that included direct fiscal stimulus spending of $1.84 trillion dollars, which is around 8.8% of GDP. This specifically focused on direct stimulus payments ($1400), state and local aid, extended unemployment benefits, vaccination funding, and education sector reopening support. Additional legislation named as Paycheck Protection Program and Health Care Enhancement Act ($483 billion) and the Coronavirus Preparedness and Response Supplemental Appropriations Act ($8.3 billion) are among the classical examples of relief packages [34].

The US Federal Reserve cut down the interest rates to near-zero and facilitated liquidity through quantitative easing strategies like asset purchase programs and repo operations. In an attempt to grant easy credit. The Central bank made several measures, such as a Main street lending program and Paycheck Protection Program liquidity facility to aid small and medium enterprises, credit facilities to stabilize bond markets. These regulatory relief programs gave temporary ease to economies in dealing with the crisis [34].

6.3.1.2 Indian Economic Measures

In India, the policy response to the pandemic included monetary, fiscal, and macro-financial measures. Initial fiscal intervention encompasses healthcare infrastructure enhancements (0.1% of GDP) and social protection including food and cash transfers (1.3% of GDP), and wage support (0.5% of GDP). Direct expenditure contributed for about 3.5% of GDP and credit-related support initiatives as 5.3% of GDP. Other than these several measures to support sectors like agriculture, urban housing, and industry with programs like the PLI scheme (Production Linked Incentive Scheme) which spend around 0.8% of GDP over 5 years [34].

In the year 2021, an amount of INR 350 billion was allocated for vaccination and for INR 800 million for free public distribution of food grains to 800 million individuals. Waivers on essential COVID-19 medical supplies were given and disaster mitigation funds for states were expedited. Support to micro, small, and medium enterprises (MSMEs), street vendors, loans and interest-free loans and subsidies and tax rebates to farmers were announced. The monetary measures by the Reserve Bank of India (RBI) in 2021 lowered its policy rates and implemented liquidity measures equivalent to 5.9% of GDP. Liquidity facilities supported rural banks, housing finance companies, and mutual funds. To ease emergency financing, emergency credit line schemes were extended to micro, small, and medium enterprises. For exchange rate stabilization, there is an increase in portfolio investment limits, forex swap, adjustments in foreign direct investments policy to safeguard critical sectors and vulnerable population of the country fighting against the pandemic [34].

The Indian economy also showed significant recovery since FY2022, after absorbing the shock from three waves of COVID-19. Some sectors are yet to regain the pre-COVID momentum, but overall growth figures and high-frequency indicators are showing a positive sign. India registered a growth of 8.7% in gross domestic product (GDP) in FY2022. This surpassed the pre-COVID level, reviving the hope for India to become the third largest economy. The policies by the government and fiscal measures taken by the government boosted the Indian economy. There were different schemes announced by the government to support the economy, such as the Pradhan Mantri (PM) Gati Shakti National Master Plan (NMP), which aimed to provide multimodal connectivity to various economic zones, and this made the movement of people, goods, and services seamless. This is often accomplished through public sector services and insurance schemes that lower out-of-pocket costs [47]. Following the devastating financial impact of COVID-19 hospitalizations during the pandemic in India, a stimulus package of INR 630,000 crore (US $7.5 billion) for FY 2022 to support the Loan Guarantee Scheme for sectors impacted by the pandemic, the Emergency Credit Line Guarantee Scheme, ₹15,000 crore for a new public health scheme, etc., were allocated. Moreover, to boost the manufacturing section, the Production-Linked Incentive (PLI) Scheme was introduced with the aim of providing INR 2 lakh crore over 5 years, starting FY2022. By monetizing central government assets, it was also aimed to raise INR 6 lakh crore over the 4 years FY 2022–2025 under the National Monetization Pipeline. As in the chapter we mentioned, the MSME sector faced the most severe losses, so to boost the sector, the central government of India took some key measures, such as the amendment of the MSME definition to remove the distinction between manufacturing and service MSME and increase the limit. Due to this increased limit, more industries will fall in the MSME category and get a boost [47]. The Reserve Bank of India fixed the repo rate at 4% in FY2022, which helped in economic recoveries and sustained growth. In order to standardize guidelines for bank and nonbank financial company loans, the RBI updated the microfinance regulations. It established guidelines for determining loan repayment caps, determining household income, and guaranteeing equitable loan pricing [47].

6.3.1.3 European Recovery Efforts

The European Union (EU) announced a number of monetary and fiscal measures to combat the economic impact of the COVID-19 pandemic. These policies prioritized the healthcare relief packages, financial stability, the problem of climate change, and social security nets to aid immediate and long-term equilibrium of the economic growth (Table 6.1).

The Next Generation EU (NGEU) recovery package was finalized and announced in December 2020 to provide fiscal relief. This package, valued at €750 billion, consists of €390 billion in grants and €360 billion in loans, financed through borrowing at the EU level. Under this initiative, the Recovery Resilience

Table 6.1 COVID-19 policy responses and recovery indicators across major economies (2020–2022)

Country/ region	Fiscal measures	Monetary measures	Recovery indicators
United States	CARES Act ($2.3T); Consolidated Appropriation Act ($868B); American Rescue Plan ($1.84T)	Near-zero interest rates; Quantitative easing; Main Street lending program	Unemployment reduced from 15% to 3.6%; Employed people reached 158.8M in Q4 2022
India	Healthcare infrastructure (0.1% GDP); Social protection (1.3% GDP); Direct expenditure (3.5% GDP); Credit support (5.3% GDP)	Policy rate reduction; Liquidity measures (5.9% GDP); Emergency credit lines for MSMEs	8.7% GDP growth in FY2022; Repo rate maintained at 4%
European Union	Next Generation EU (€750B); SURE program (€100B); Pandemic Crisis Support (€240B	Pandemic Emergency Purchase Programme (€1,350B); Extended credit lines	Recovery and Resilience Facility implementation; 37% budget allocated to European Green Deal

Facility (RRF) is made to channel the funds in 2021–2023, 70% of the grants were allocated for the year 2021–2022 where the impact of COVID was severe. European countries having high debt like Spain and Italy and other eastern European nations gained the most from these relief packages [34]. This initiative represented a paradigm shift in economic policy-making, linking recovery efforts to green transitions, digital transformation, and social cohesion [5].

The EU also introduced several other fiscal safety nets to provide targeted relief. The Pandemic Crisis Support (PCS) was also introduced in 2021 offering €240 billion in credit lines for euro-area countries to finance health-related spending, A €100 billion temporary loan instrument aimed at protecting workers and jobs via the SURE program. Also, the EU expanded loan guarantees through the European Investment Bank (EIB), which used €25 billion in guarantees to support small and medium-sized enterprises (SMEs). The EU's Stability and Growth Pact was also activated and extended through 2022, which enabled member states to increase spending without breaching deficit limits [34].

On the monetary side, the European Central Bank established the €1350 billion Pandemic Emergency Purchase Programme (PEPP) in March 2020 to provide financial assistance to member states that were experiencing financial strain during the pandemic. The EU has come up with its plan for the multiannual financial framework while drafting the recovery plan that it called "Next Generation EU" using the aid of an extra 750 billion euros. That plan was accepted in December 2020. It allowed deficit spending through mutual bonds and fiscal transfers for the first time in the EU. The package further has RRF resources equivalent to €672.5 billion that should help recover the Member States from this pandemic. At least 37% of the budget is put into supporting the EGD with its objective of getting the EU carbon neutral by 2050. It permitted deficit financing through shared bonds and fiscal transfers, which was a first for the EU. The plan includes €672.5 billion for the Recovery and Resilience Facility (RRF), which aims to help member states recover from the pandemic. At least 37% of this fund is dedicated to supporting the European Green Deal (EGD), with the goal of making the EU carbon neutral by 2050 [53].

6.3.2 International Cooperation and Governance

In order to combat the COVID-19 pandemic, wealthy countries have undertaken extensive public spending and health initiatives. On the other hand, less wealthy nations frequently depend on assistance from international organizations and have fewer financial and fiscal options [48].

Therefore, a number of significant financial institutions and regional organizations raised a significant amount of money to aid LMICs. As an early responder, the World Bank pledged US $160 billion over 15 months, including a US $12 billion fast-track package intended to improve COVID-19 response and hasten recovery in LMICs through grants and low-interest loans [23]. The United Nations Development Program, UNDP, has a network of 3100 partners that respond to socio-economic issues. They have a budget of US $500 million, strategically divided into three critical areas: $150 million was allotted for health system support, $250 million for crisis management, and $100 million for socio-economic impact assessment [23].

The European Union came out as the largest international donor and contributed almost 57% of the global development assistance even though its share in the global economy is only one-fifth. This response package prioritizes the vulnerable economies and nations, like the African regions (EUR 3.25 billion) to uplift their development. These initiatives were further supported by the creative "Team Europe" initiative, which brought together resources from EU Member States, financial institutions, and existing programs to create a package worth EUR 20 billion [36].

Two significant financial institutions in Asia took the initiative. In addition to the US $4.6 billion already set aside for particular projects, the Asian Development Bank (ADB) committed US $20 billion in loans, grants, and technical assistance. They provided significant funding (US $1.5 billion each) to Indonesia, the Philippines, and India, with an emphasis on both health initiatives and mitigating the effects on the economy [57]. Meanwhile, a US $5 billion crisis emergency fund was announced by the Asian Infrastructure Investment Bank (AIIB) to facilitate the recovery. This move showed China's prominent role in the regional response of COVID-19 crisis [13].

These combined and coordinated efforts by all the major economies and financial institutions and development banks showed ways to combat this health emergency in the first place, and then in combating the economic and social burns that the coronavirus pandemic brought in front of the whole world. The focused strategies targeted the lower income, and the most vulnerable population that were most affected, both in terms of lives and livelihood due to the pandemic.

In 2021, official development assistance (ODA) from OECD countries hit a record US $179 billion, with real terms increasing by 4.4% compared to 2020. This surge was primarily attributed to the global response to the COVID-19 pandemic. Approximately US $6.3 billion, or 857 million doses, of the total ODA was used to donate COVID-19 vaccines. Of these, US $2.3 billion was spent on surplus domestic doses, US $3.5 billion in buying vaccines specifically for developing countries, and US $0.5 billion on associated costs. Spending on COVID-19 activities was US $18.7 billion of the total ODA, or 10.5%, slightly higher than 10.2% in 2020 [49].

The US $18.8 billion allotted to humanitarian aid, a 3.5% increase from 2020, helped to support wider humanitarian needs as well. Debt relief, however, remained minimal at US $545 million, and aid for refugees hosted in donor countries stood at US $9.3 billion, relatively stable from the previous year. The largest share of ODA disbursements continued to be grants. Loans and equity investments dropped to 19% of bilateral ODA from 21% in 2020. The largest share of loans was held by Japan, Korea, and France [49]. The largest increases were recorded in ODA from donor countries such as Italy, Korea, and Ireland. Five nations—Denmark, Germany, Luxembourg, Norway, and Sweden—met or exceeded the UN target of 0.7% ODA to GNI [49].

While ODA has increased steadily over decades, it is in times of crisis that its critical role really shines through. The OECD calls for continued efforts aimed at strengthening the vulnerabilities of developing countries, especially considering that multiple global challenges such as food insecurity and geopolitical tensions augment the pressures on aid systems [49].

6.4 Future Preparedness and Digital Transformation

The COVID-19 crisis has posed a number of challenges in all spheres of life. And policymakers and governments are still finding ways to overcome its aftereffects. There are new innovations and technological developments happening in the healthcare systems, education sector, and the digital economy is shaping like never before. Remote work gained immense prominence after the pandemic and it gave more flexibility to both employees and employers. Small and even established firms created their online platforms and marketplaces to adapt to the new normal thus expanding e-commerce industries. Digital transformation is not a smooth process, rather it has its own difficulties, like the technological know-how and gap in Research and Development, challenges in adoption of ICT in different sectors [38]. Furthermore, the crisis has established new frameworks for future preparedness, emphasizing the need for flexible governance structures, rapid response mechanisms, and strong international coordination protocols. Successful recovery strategies have shown the importance of balancing immediate crisis response with long-term resilience building, particularly in strengthening healthcare infrastructure, social safety nets, and public-private partnerships. The role of technology and digital transformation has emerged as particularly crucial, transcending its traditional role as a business continuity tool to become a fundamental component of economic resilience.

6.5 Conclusion

The COVID-19 pandemic showed how over-reliance on one or two global supply chain units is a risky thing for an economy. When the production and supply chains were closed in China because of the pandemic it impacted all the interconnected economies and caused instability and huge shortages. It led to a push for decentralized economic systems, where individual economies diversified their local production units, and it also helped them understand near-shoring to reduce dependency on global supply chains. This shift shows that businesses now care more about being prepared for disruptions rather than saving money and being efficient in the short term.

The coronavirus crisis also emphasized the strengths and weaknesses of working together. In the first wave of COVID-19, Countries were tackling the spread of the pandemic by focusing on their own needs only. However, in today's interconnected network of economies, it is practically impossible for a single country to work in isolation and combat such an unprecedented crisis alone. Therefore in the later phases or as the pandemic hit the second and third waves, international organizations, healthcare providers, and Governments realized the importance of coordinated responses and tried to develop approaches that could help in international cooperation and working together as one single unit globally. This cooperation helped in the vaccine trials, maintaining social distancing and international travel protocols, and encouraging and supporting each other throughout the pandemic. This reveals how international Corporations can help address the global challenges efficiently. As evident in the pandemic, the advanced economies made a pathway for the weaker countries, and international organizations like the IMF, World Bank, Asian Development Bank, etc., helped the vulnerable economies of the world to deal with the crisis. However, this also depicted a critical gap in international coordination and cooperation among countries. Some economies found it hard to get the aid and adequate resources to prevent the spread of this public health emergency.

The crisis also gave several lessons for future preparedness and crisis management. First, the pandemic made it clear how important it is to provide social safety nets like public food, distribution, programs, employment benefits, etc., for the economically weaker sections of society. There is no doubt that a strong and stable fiscal and monetary space is essential when the economy is in a boom period so that it can provide quick response and essential support to the population at times of emergencies, natural disasters, and uncertain shocks.

As it was evident this crisis was different from the earlier outbreaks of diseases or financial recession in terms of simultaneous demand and supply side strains, and digital preparedness. Today in the twenty-first century we live in a state of rapid technological development. Digital transformation has taken a new role and shape during and after this pandemic. Nations who had a better digital infrastructure were quick to analyze and recover from the shocks of the pandemic as they restructured and transformed their economies digitally and shifted employment patterns in remote works whereas, countries with poor infrastructure and lack of advanced technology found it hard to adopt the newer changes. This also shows the gap or digital divide that the world is having today. Even within a nation, there are parts of metropolitan cities with fast availability of internet and 5G services. On the parallel end, are villages and rural areas with no mobile networks and internet services and with poor digital literacy in people to overcome the large-scale impact of the crisis.

Good governance also played a prominent role in crisis management. Countries with sound political, economic, and public healthcare systems were in a better position to tackle the challenges as compared to nations that have weak institutional capacity and public trust in them. Honest and clear Communication as well as providing assurance by the government and policymakers to its citizens is equally important in such an emergency situation where there is fear and panic among everyone.

The pandemic has left an indelible mark on human society and the economy globally.

Challenges are huge and not yet solved completely. The global economy is still recovering from its shocks and trying to adapt itself to what is called the "new normal". In an attempt to mitigate the crisis, the pandemic pushed innovation, technology, and digitalization for economies. The education sector has undergone a shift from traditional to virtual classrooms. The healthcare sector has also moved towards remote collaborations between patients and healthcare providers for diagnosis, and treatment. Artificially intelligent models are also developed for the examination and provision of healthcare services. The need for more inclusive and sustainable development practices is there as the pandemic has pushed more population below the poverty line and lack access to a clean and healthy environment. These lessons of collective efforts, long-term transformation in the system, prioritizing resilience, and enhancing the institutional economic capacities will serve as a guide for future uncertain crisis preparedness.

References

1. Abbas J, et al. Exploring the impact of COVID-19 on tourism: transformational potential and implications for a sustainable recovery of the travel and leisure industry. Curr Res Behav Sci. 2021;2(100033):100033.
2. Abdelrahman Z, Li M, Wang X. Comparative review of SARS-CoV-2, SARS-CoV, MERS-CoV, and influenza A respiratory viruses. Front Immunol. 2020;11:552909.
3. Aditya B, Amri I. Rethinking informal economy resilience during crisis: experience from COVID-19 pandemic. Indian J Labour Econ. 2023;66(3):711–37.
4. Alberola-Ila E, et al. The fiscal response to the Covid-19 crisis in advanced and emerging market economies. 2020. Available at: https://www.bis.org/publ/bisbull23.htm. Accessed 14 Dec 2024.
5. Alcidi C, Gros D. Next generation EU: a large common response to the COVID-19 crisis. Inter Econ. 2020;55(4):202–3.
6. Baek S, Mohanty SK, Glambosky M. COVID-19 and stock market volatility: an industry level analysis. Financ Res Lett. 2020;37(101748):101748.
7. Baig AS, et al. Deaths, panic, lockdowns and US equity markets: the case of COVID-19 pandemic. Financ Res Lett. 2021;38(101701):101701.
8. Bakar NA, Rosbi S. Effect of coronavirus disease (COVID-19) to tourism industry. Int J Adv Eng Res Sci. 2020;7(4):189–93.
9. Barman A, Das R, De PK. Impact of COVID-19 in food supply chain: disruptions and recovery strategy. Curr Res Behav Sci. 2021;2(100017):100017.
10. Borio C. The Covid-19 economic crisis: dangerously unique. Bus Econ (Cleveland, Ohio). 2020;55(4):181–90.
11. Teodorović J. Global health and safety trends in tourism. Hotellink. 2020;1:58–65.
12. Bruinen de Bruin Y, et al. Initial impacts of global risk mitigation measures taken during the combatting of the COVID-19 pandemic. Saf Sci. 2020;128(104773):104773.
13. Career Opportunities. AIIB looks to launch USD5 billion COVID-19 crisis recovery facility. n.d. Available at: https://www.aiib.org/en/news-events/news/2020/AIIB-Looks-to-Launch-USD5-Billion-COVID-19-Crisis-Recovery-Facility.html. Accessed 16 Dec 2024.
14. Chan JF-W, et al. A familial cluster of pneumonia associated with the 2019 novel coronavirus indicating person-to-person transmission: a study of a family cluster. Lancet. 2020;395(10223):514–23.
15. Chaudhary R, Bakhshi P, Gupta H. Volatility in international stock markets: an empirical study during COVID-19. J Risk Financ Manag. 2020;13(9):208.
16. Chauhan N, et al. Rise and fall of Covid–19 cases in India: a review. n.d. Available at: https://www.jebmh.com/articles/rise-and-fall-of-covid%2D%2D19-cases-in-india-a-review.pdf
17. Ciotti M, et al. The COVID-19 pandemic. Crit Rev Clin Lab Sci. 2020;57(6):365–88.
18. Crisp. Crisp: the collaborative commerce platform. Crisp; 2022. Available at: https://www.gocrisp.com/demandwatch. Accessed 14 Dec 2024.
19. CRS Report. 2021. Available at: https://crsreports.congress.gov/product/pdf/R/R46554/9. Accessed 14 Dec 2024.
20. Cutler DM. Health system change in the wake of COVID-19. JAMA Health Forum. 2023;4(10):e234355.
21. Deshpande A. The Covid-19 pandemic and gendered division of paid work, domestic chores and leisure: evidence from India's first wave. Econ Politica. 2022;39(1):75–100.
22. Di Fabio A. The psychology of sustainability and sustainable development for well-being in organizations. Front Psychol. 2017;8:1534.
23. Ehrenberg JP, et al. Efforts to mitigate the economic impact of the COVID-19 pandemic: potential entry points for neglected tropical diseases. Infect Dis Poverty. 2021;10(1):2.
24. El Keshky MES, Basyouni SS, Al Sabban AM. Getting through COVID-19: the pandemic's impact on the psychology of sustainability, quality of life, and the global economy—a systematic review. Front Psychol. 2020;11:585897.

25. Espinosa-Méndez C, Arias J. COVID-19 effect on herding behaviour in European capital markets. Financ Res Lett. 2021;38(101787):101787.
26. Essien LS, Levinstein MD, Owens G. Unemployment rate returned to its prepandemic level in 2022. Bureau of Labor Statistics; 2023. Available at: https://www.bls.gov/opub/mlr/2023/article/unemployment-rate-returned-to-its-prepandemic-level-in-2022.htm. Accessed 14 Dec 2024.
27. Filip R, et al. Global challenges to public health care systems during the COVID-19 pandemic: a review of pandemic measures and problems. J Pers Med. 2022;12(8):1295.
28. George TK, et al. The economic impact of a COVID-19 illness from the perspective of families seeking care in a private hospital in India. Dialogues Health. 2023;2(100139):100139.
29. Gursoy D, Chi CG. Effects of COVID-19 pandemic on hospitality industry: review of the current situations and a research agenda. J Hosp Mark Manag. 2020;29(5):527–9.
30. Haakenstad A, et al. Catastrophic health expenditure during the COVID-19 pandemic in five countries: a time-series analysis. Lancet Glob Health. 2023;11(10):e1629–39.
31. Holshue ML, et al. First case of 2019 novel Coronavirus in the United States. N Engl J Med. 2020;382(10):929–36.
32. Hoshikawa T, Yoshimi T. The effect of the COVID-19 pandemic on South Korea's stock market and exchange rate. Dev Econ. 2021;59(2):206–22.
33. ILO. 2020. Available at: https://www.ilo.org/wcmsp5/groups/public/%2D%2D-dgreports/%2D%2D-dcomm/documents/briefingnote/wcms_740877.pdf. Accessed 14 Dec 2024.
34. IMF. Policy responses to COVID19. IMF; 2020. Available at: https://www.imf.org/en/Topics/imf-and-covid19/Policy-Responses-to-COVID-19#E. Accessed 15 Dec 2024.
35. Jomo KS, Chowdhury A. COVID-19 pandemic recession and recovery. Soc Int Dev. 2020;63(2–4):226–37.
36. Josep Borell. "Team Europe"—global EU response to Covid-19 supporting partner countries and fragile populations. EEAS; 2020. Available at: https://www.eeas.europa.eu/eeas/%E2%80%9Cteam-europe%E2%80%9D-global-eu-response-covid-19-supporting-partner-countries-and-fragile-populations_en. Accessed 16 Dec 2024.
37. Kalyani M. Indian informal sector: an analysis. n.d. Available at: https://www.arcjournals.org/pdfs/ijmsr/v4-i1/9.pdf. Accessed 15 Dec 2024.
38. Kutnjak A. Covid-19 accelerates digital transformation in industries: challenges, issues, barriers and problems in transformation. IEEE Access. 2021;9:79373–88.
39. Lambovska M, Sardinha B, Belas J Jr. Impact of the covid-19 pandemic on youth unemployment in the European Union. Econ Manag Spect. 2021;15(1):55–63.
40. Le TDQ, et al. Government intervention and stock price returns during covid-19 pandemic: evidence from an emerging market. Cogent Bus Manag. 2024;11(1). Available at: https://www.tandfonline.com/doi/pdf/10.1080/23311975.2024.2376042
41. Li Z, et al. A comparative analysis of COVID-19 and global financial crises: evidence from US economy. Econ Res. 2022;35(1):2427–41.
42. Magesh. View of effect of COVID-19 on unemployment in India. 2023. Available at: https://www.shanlaxjournals.in/journals/index.php/economics/article/view/6218/5826. Accessed 15 Dec 2024.
43. Magesh Kumar C, Sujatha K, Rajesh Kumar K. Effect of COVID-19 on unemployment in India. Shanlax Int J Econom. 2023;11(3):51–5.
44. Mazur M, Dang M, Vega M. COVID-19 and march 2020 stock market crash. Evidence from S&P1500. SSRN Electron J. 2020. Available at: https://doi.org/10.2139/ssrn.3586603.
45. Meier M, Pinto E. COVID-19 supply chain disruptions. Eur Econ Rev. 2024;162(104674):104674.
46. Moosavi J, Fathollahi-Fard AM, Dulebenets MA. Supply chain disruption during the COVID-19 pandemic: recognizing potential disruption management strategies. Int J Disaster Risk Reduct. 2022;75(102983):102983.
47. NABARD Annual Report. NABARD annual report 2022. 2022. Available at: https://www.nabard.org/pdf/economy-in-recovery-mode-eng.pdf
48. Naseer S, et al. COVID-19 outbreak: impact on global economy. Front Public Health. 2022;10:1009393.
49. OECD. COVID-19 assistance to developing countries lifts foreign aid in 2021. OECD; 2021. Available at: https://www.oecd.org/en/about/news/press-releases/2022/04/covid-19-assistance-to-developing-countries-lifts-foreign-aid-in-2021-oecd-.html#:~:text=12%2F04%2F2022%20%2D%20Foreign,data%20collected%20by%20the%20OECD. Accessed 15 Dec 2024.
50. Pujawan IN, Bah AU. Supply chains under COVID-19 disruptions: literature review and research agenda. Supply Chain Forum Int J. 2022;23(1):81–95.
51. Rahmouni L. The impact of COVID-19 on the cinema industry. ELWAHAT J Res Stud. 2023;16(1):1084–99.
52. Raman MS, et al. Covid-19 pandemic crisis on labour and employment force in India: impact, approaches and future perspective. Indian Res J Ext Edu. 2021;21(2&3):43–52.
53. Ryner JM. Silent revolution/passive revolution: Europe's COVID-19 recovery plan and green deal. Globalizations. 2023;20(4):628–43.
54. Sahoo PK, et al. Urban to rural COVID-19 progression in India: the role of massive migration and the challenge to India's traditional labour force policies. Int J Health Plann Manag. 2022;37(1):528–35.

55. Sahu TN, et al. Female migrant workers in conflict with life and livelihood amid COVID-19: violence and gendered crisis. Int J Community Soc Dev. 2024;6(4):379–98.
56. Sarnot R. Impact of COIVD-19 on Indian economy. Page. 8(2):2022.
57. Sato A, Aboneaaj R, Morris S. How effectively is the Asian development bank responding to COVID-19? An early assessment. n.d. Available at: https://www.cgdev.org/sites/default/files/how-effectively-asian-development-bank-responding-covid-19-early-assessment.pdf. Accessed 15 Dec 2024.
58. Shafi M, Liu J, Ren W. Impact of COVID-19 pandemic on micro, small, and medium-sized enterprises operating in Pakistan. Res Glob. 2020;2(100018):100018.
59. Spiteri G, et al. First cases of coronavirus disease 2019 (COVID-19) in the WHO European Region, 24 January to 21 February 2020. Euro Surveillance. 2020;25(9). Available at: https://www.eurosurveillance.org/content/10.2807/1560-7917.ES.2020.25.9.2000178?crawler=true
60. Wildman SS. 28 new media and the need for new analytical frameworks: dual challenges to media economics and policy analysis. In: Management and economics of communication. De Gruyter; 2020. p. 485–96.
61. World Bank Group. Chapter 1. The economic impacts of the COVID-19 crisis. World Bank Group; 2022. Available at: https://www.worldbank.org/en/publication/wdr2022/brief/chapter-1-introduction-the-economic-impacts-of-the-covid-19-crisis. Accessed 15 Dec 2024.
62. Xiang S, et al. The effect of COVID-19 pandemic on service sector sustainability and growth. Front Psychol. 2021;12:633597.
63. Xing L, et al. Substitution effect of Asian economies on China's industrial and supply chains: from the perspective of global production network. Humanit Soc Sci Commun. 2024;11(1):1–27.
64. Xu Y, Xu W. The impact of COVID19 on stock market: a study of stock market interdependency. In: Advances in economics, business and management research. Dordrecht: Atlantis Press International BV; 2022. p. 1859–63.
65. Yan K, Yan H, Gupta R. The predicted trend of COVID-19 in the United States of America under the policy of "Opening Up America Again". Infect Dis Model. 2021;6:766–81.
66. Yu KDS, Aviso KB. Modelling the economic impact and ripple effects of disease outbreaks. Process Integr Optim Sustain. 2020;4(2):183–6.
67. Zaremba A, et al. Infected markets: novel coronavirus, government interventions, and stock return volatility around the globe. Financ Res Lett. 2020;35(101597):101597.
68. Zhang D, Hu M, Ji Q. Financial markets under the global pandemic of COVID-19. Financ Res Lett. 2020;36(101528):101528.

The Macroeconomic Implications of COVID-19 Economic Crisis Worldwide

7

Qazi Ghazala Mateen, Supriya Raghav, Santosh Kumar, Ankita Raghav, Virendra Kumar, and Poonam Khanna

7.1 Introduction

Throughout history, the world has witnessed many infectious disease outbreaks that have affected mortality rates and led to the collapse of entire economic systems. Examples of such diseases include The Spanish flu of 1918, the Ebola outbreak of the twenty-first century, and the recent coronavirus pandemic. While the earlier outbreaks were severe and devastating, none had shaken the interconnected global economy as COVID-19.

COVID-19 was caused by the novel coronavirus SAR-coV-2 which was originated in the city Wuhan, China, and in a short span of time this virus spread and escalated into an epidemic forcing the World Health Organization (WHO) to declare it a global pandemic on March 11, 2020 [7]. Initially, the coronavirus pandemic was a public health emergency worldwide arising out of non-economic reasons. At the end of the year 2021, more than 287 million confirmed cases of COVID-19 across the world had been reported to the World Health Organization (WHO) which includes 5.4 million confirmed deaths [51]. This exposed the shortcomings in global health systems and inadequate government preparedness to manage such multifaceted and uncertain challenges. The original public health emergency soon turned into an economic crisis even deeper than the great depression of the 1930s and the oil shocks of the 1970s.

The COVID-19 pandemic has had major macroeconomic repercussions. It impacted both the demand and supply side of the economy simultaneously [6]. Supply was particularly impacted due to preventive steps like strict lockdowns, social distancing protocols, and restricted mobility within and outside the borders were implemented to control the disease spread. There were sharp contractions both in production and consumption. As a result of the sudden global shutdown, firms and labor markets hit hard. Companies did a mass firing of employees. A large number of the population worldwide lost their jobs and daily bread and butter. Those working in informal sectors and semi-skilled workers were severely impacted. Forecasts for Global GDP plummeted, with service-driven and tourism-dependent economies experiencing significant challenges. On the demand side, the rise in inflation and unemployment shifted the

Q. G. Mateen
Department of Economics, Aligarh Muslim University, Aligarh, India

S. Raghav (✉) · S. Kumar
School of Public Health, Poornima University, Jaipur, India

A. Raghav · V. Kumar
Department of Economics, IIMT University, Meerut, India

P. Khanna
School of Public Health, Post Graduate Institute of Medical Education and Research, Chandigarh, India

A. Raghav et al. (eds.), *COVID-19 Impact on Economy, Environment and Healthcare*,
https://doi.org/10.1007/978-981-95-1196-9_7

consumer behavior and spending patterns. There was a fall in aggregate demand and hence it changed the investment patterns too. This dual shock triggered a synchronized global recession, with record economic contractions reported across regions [6]. As a consequence, 93 million more people worldwide were pushed into extreme poverty post pandemic [58].

To navigate through the COVID-19 pandemic, policymakers and monetary authorities around the world announced huge amounts of fiscal stimulus and monetary measures. As governments applied deficit-spending fiscal strategies to stimulate aggregate demand, the Keynesian economic school has gained renewed relevance [8]. Keynesian macroeconomic framework promotes economic stimulus through fiscal policy intervention tax cuts and increase in state spending to boost the aggregate demand. While these measures mirrored past crises responses, they were unparalleled in scale and scope, reflecting the unique challenges posed by the pandemic.

7.2 Global Economic Consequences and Dynamics of Growth

The pandemic had caused severe contractions in all output-generating sectors, which in turn impacted the production dynamics for firms and industries and had a ripple effect on economic growth, causing widespread downturns and disruptions in trade and commerce.

Various forecasts suggest that GDP will shrink by a median of around 6.5% during 2020, with countries experiencing 10% or even 15% contractions [19, 54]. The expectation of global economic growth has been revised by the IMF from 3.4% to 4.9% for the year 2020 [25]. This shrink in growth was particularly more in European economies like Portugal, Greece, and Spain with large service sectors, such as hospitality and tourism, and had cascading effects through international supply chains [19, 38, 43]. Also, there has been a significant decrease in employment and incomes globally as studies estimated a loss of 147 million full-time equivalent jobs and an astonishing $2.1 trillion in economic damage [43]. This section aims to examine the scale and shape of these economic events, focusing on global macroeconomic consequences, output contractions, and emergent growth patterns.

7.2.1 Output Loss and Growth Trajectory

The pandemic has induced multifaceted supply chain disruptions causing an increase in the cost of trade thereby hindering both import and export volumes and resulting in significant output loss [48, 56]. Without a timely implemented policy intervention, the pandemic is likely going to cause a major, long-lasting recession. The effects may be deeper and more long-lasting for advanced economies, including those in the United Kingdom, compared to the emerging Asian economies [12].

7.2.1.1 Economy-Level Analysis

The data on global GDP growth depicts striking changes in the GDP of all economies in the years after COVID-19 as depicted in Table 7.1. There was a recession in 2020 as the world's GDP dropped to −3.1% with developed markets shrinking by −4.5%, and developing economies contracted by −2.1%. This was succeeded by a vigorous growth spurt in 2021 where global growth reached 6% due to emerging markets performing at 6.6%, with developed countries contributing at 5.2%. The period of 2023–2024 reports suggest that the growth rates across all the economies would rather revert to a steady growth with the growth of the world economy settled at 2.9–3.0%. More advanced economies are settling at lower rates: 1.4–1.5%, while latest emerging markets still demonstrate higher growth rates, which are 4.0%.

7.2.1.2 Country-Level Analysis

By looking at the national-level data statistics, significant differences in the effect and recovery patterns for different economies could be obtained as shown in Table 7.2. The United States, Japan, and Germany all were hit by nega-

Table 7.1 Annual real GDP growth (percent change)

Economy group	2019	2020	2021	2022	2023	2024
World	2.8	−3.1	6.0	3.5	3.0	2.9
Advanced economies	1.7	−4.5	5.2	2.6	1.5	1.4
Emerging market and developing economies	3.7	−2.1	6.6	4.1	4.0	4.0

Source: International Monetary Fund, World Economic Outlook Database [28–33]

Table 7.2 Major economies performance across regions (real GDP growth, %)

Country/ Economy type	2019	2020	2021	2022	2023	2024
Advanced economies						
United States	2.2	−3.4	5.7	2.1	2.1	1.5
Germany	0.6	−4.6	2.6	1.8	−0.5	0.9
Japan	0.7	−4.6	1.7	1.0	2.0	1.0
Emerging market economies						
China	6.1	2.3	8.1	3.0	5.0	4.2
India[a]	4.2	−7.3	8.7	7.2	6.3	6.3
Brazil	1.1	−4.1	4.6	2.9	3.1	1.5

Source: IMF World Economic Outlook Database [28–33]
[a]Note: For India, data is presented on a fiscal year basis (2019–2020, 2020–2021, 2021–2022, 2022–2023, 2023–2024)

tive growth in the pandemic year but the United States recovery pattern showed an instant recovery trend rising to 5.7% GDP growth compared to −3.7% in the previous year. Japan and Germany both had −4.6% annual growth in 2020 and Germany's recovery pattern is uneven as it got positive in 2021 and 2022 but falling back to −0.5 in 2023. Japan remained somewhere between 1.0 and 2.0 throughout its recovery phases. In emerging economies, China was the only country to maintain positive output growth in 2020 and it expanded strongly at 8.1% in 2021. However post 2021 China's growth also slowed down and remained between 3.0% and 5.0%. India, despite going through the most critical initial contraction of −7.3%, managed to recover and grow at 8.7% in 2021. Brazil's growth rates stabilized after a significant contraction but its forecasted growth for 2023–2024 continued to decline and its recovery pattern is slow.

7.2.1.3 Comparative Analysis and Way Forward

A review of the two tables reveals several key trends and insights about the post-COVID-19 global economic landscape. First, there is a clear divergence between developed and developing economies, both in their initial resilience to the crisis and their recovery models. While advanced economies proved more vulnerable to the initial shock, emerging markets, especially in Asia, showed potential recovery and stronger growth prospects.

National-level data confirm these general trends, while highlighting important regional differences. The Asian economy (China and India) always exceeds the market friends and suggests a constant change in economic gravity related to Asia. Among developed countries, the United States has demonstrated more economic stability compared to Europe and Asian similarities. This indicates a variety of economic adaptability in the "advanced economy" category.

Looking ahead to 2024, the data points to three important conclusions: (1) the global economy is moving toward a "new normal" with stable but differentiated growth rates; (2) the growth premium of emerging markets relative to developed countries is expected to persist, reflecting structural advantages in terms of demographics and development potential; (3) regional economic disparities, particularly the strength of Asian economies, are likely to widen. These trends suggest that the COVID-19 crisis may accelerate existing patterns of rebalancing the global economy, with long-term implications for international economic relations and development patterns. This comprehensive analysis of global economic growth patterns provides a basis for understanding employment patterns and subsequent changes in the labor market that will be examined in the next section of this chapter.

7.3 Employment Patterns and Labor Markets: Shocks and Adjustments Amid the COVID-19 Economic Crisis

The COVID-19 pandemic has instigated unparalleled upheavals within global labor markets, culminating in notable economic shocks and requisite adjustments. This synthesis investigates the ramifications on employment patterns, the disparate effects experienced by various worker demographics, and the reactive measures undertaken by labor markets across diverse geographical regions.

Within a few months of the COVID-19 pandemic, labor markets were pushed into a deep recession, causing a rise in unemployment rates. In the United States, the rate of unemployment rose to about 20% in April 2020, with significant job losses recorded in all sectors and states [4]. In a parallel manner, the European Union (EU) encountered significant fluctuations in employment figures, albeit the repercussions were less severe than those observed in the United States [36]. In developing nations, high-frequency telephone surveys revealed considerable work disruptions, particularly among female, young, less educated, and urban worker cohorts [41].

The dataset shown in Table 7.3 presents international unemployment trends during the peak of the COVID-19 crisis, There is a notable global rise in unemployment, as the global unemployment rate increases from 5.6% in 2019 to 6.6% in 2020. This one percentage point rise is equal to 38 million employees globally who became unemployed in the pandemic year as the total number of unemployed people increased from 186 million in 2019 to 224 million in 2020.

The effect of the pandemic on unemployment was inconsistent across different economies. Lower-middle income countries faced the most severe percentage point rise in unemployment rates, going from 5.9% to 6.9%, which implies 17 million additional unemployed individuals. High-income countries, although beginning with a lower base rate of 4.8% in 2019, also witnessed a rise to 6.5% in 2020, equating to 11 million more unemployed people. Upper-middle income countries, despite starting with a comparatively high initial unemployment rate of 5.8%, saw a somewhat less pronounced rise to 6.5% in 2020, yet this still resulted in 8 million additional unemployed individuals. Interestingly, low-income countries seemed to experience the smallest rise in unemployment rates, from 5% to 5.7%, resulting in an addition of 2 million unemployed individuals.

Nevertheless, this relatively small rise may be linked to the prevalence of informal employment in these economies, which frequently remains unrecorded in official unemployment data. These statistics highlight the unparalleled scale of the COVID-19 pandemic's effect on global labor markets, as all economic classifications faced substantial increases in unemployment rates and total numbers of unemployed individuals. The overall increase of 38 million unemployed individuals around the world in a single year represents one of the most severe global labor market disruptions in contemporary history, emphasizing the extraordinary nature of the COVID-19 economic crisis.

Table 7.3 Total unemployment rates for population (15+) peak COVID-19

	Unemployment rate (percentage)	Unemployment rate (percentages)	Unemployment (millions)	Unemployment (millions)
Economy type	2019	2020	2019	2020
WORLD	5.6	6.6	186	224
Low-income countries	5	5.7	12	14
Lower-middle income countries	5.9	6.9	64	81
Upper-middle income countries	5.8	6.5	80	88
High-income countries	4.8	6.5	29	40

Source: ILOSTAT, ILO modelled estimates [26]

The years post the peak of the pandemic indicate a consistent recovery in global labor markets. By 2021, unemployment figures started displaying improvement with 6.2% in 2021 globally, continuing to reduce to 5.9 and 5.7 million in 2022 and 2023, respectively. The recent ILO stats shows it stays at 4.9 for the total world economy in 2024. High-income nations exhibited particularly robust resilience, with unemployment rates gradually declining to 4.7% by 2023. Lower-middle-income nations also exhibited encouraging progress, reducing their unemployment rate to 5.4%. However, the recovery path has posed more challenges for upper-middle-income nations, where unemployment rates persisted relatively high at 6.3% by 2023, indicating a slower pace of recovery. These differing trends across economic groups underscore the intricate nature of post-pandemic labor market dynamics and suggest that while the overall trend is positive, ongoing policy focus may be required to guarantee sustained recovery across all economic segments.

7.3.1 Gender and Employment Patterns Amid Pandemics

The COVID-19 pandemic has caused major disruptions to labor markets around the world, with impacts that are markedly different from previous economic crises. Previous recessions, such as the 2008 financial crisis, primarily affected male employment and were referred to as "mancession" [5]. The pandemic, on the other hand, has triggered what economists call a "shecession," resulting in a clear shift in how economic shocks affect gender-based employment patterns.

The 2020 unemployment statistics reveal intriguing trends across various economies, underscoring the distinct gender-based consequences of the pandemic (Table 7.4). The data illustrates three unique patterns across these economies. Initially, in developed economies, the gender gap in unemployment was relatively minor. For example, the United States displayed a slight difference with male unemployment at 15% and female at 14.8%, while Germany exhibited a negligible gap with female unemployment at 8%, just above male unemployment at 7.7%. Japan recorded the lowest unemployment rates for both genders, with male unemployment at 5.3% and female at 4.1%.

Conversely, emerging economies presented a markedly different scenario. Brazil revealed the most significant disparity with female unemployment markedly higher at 35.3% compared to male unemployment at 26.3%, representing a considerable gap of 9 percentage points. In a similar vein, India showcased significant gender disparity with female unemployment at 25.5% in contrast to male at 24.3%. The United Arab Emirates also displayed a pronounced gap with female unemployment at 18.7%, juxtaposed with male unemployment at 12.5%.

Various factors may elucidate these disparities. In emerging economies, the elevated gender gaps

Table 7.4 Annual unemployment rate percentage by sex and age (15–54) (2020)

Country	Unemployment rate (Male)	Unemployment rate (Female)	Source
United States of America	15	14.8	LFS—Current Population Survey
United Kingdom of Great Britain and Northern Ireland	15.3	11.8	LFS—Labor Force Survey
Germany	7.7	8	LFS—EU Labor Force Survey
Canada	20.7	19.4	LFS—Labor Force Survey
Japan	5.3	4.1	LFS—Labor Force Survey
United Arab Emirates	12.5	18.7	LFS—Labor Force Survey
India	24.3	25.5	LFS—Periodic Labor Force Survey
Brazil	26.3	35.3	HS—Continuous National Household Sample Survey

Source: ILOSTAT [26]

may be ascribed to the predominance of women employed in informal sectors and service industries that were more adversely affected by lockdown measures. Moreover, the responsibility of unpaid care work, especially considering school closures and remote learning, disproportionately burdened women. Cultural factors and entrenched gender roles within these societies may have further influenced women's capabilities to retain or re-establish employment during the crisis. The relatively narrower gaps in developed economies can be linked to more robust social safety nets, more formalized employment structures, and enhanced access to remote work facilities, which served to mitigate the pandemic's effects on female employment. This diverse impact of the pandemic on gender-specific employment was nevertheless relatively brief in many economies.

Research conducted across 38 advanced and emerging market economies indicated that although 53–68% of nations experienced she-cessions during the peak of the crisis in the second quarter of 2020, approximately two-thirds of the affected nations experienced a reversal of this gender gap by the third quarter [5]. The differing degrees of recovery between advanced and emerging economies underscored the vital influence of policy measures and economic frameworks in shaping the pandemic's repercussions on gender-based employment patterns. Nevertheless, the implications of these temporary disruptions may yield more enduring effects on gender equality within the workforce. The pandemic revealed and potentially intensified existing vulnerabilities in women's employment, particularly in emerging economies. This indicates a necessitation for targeted policies that address not only immediate employment challenges but also bolster the resilience of female employment against prospective economic shocks.

7.3.2 Analysis of Trends of Remote Work and Changes in Employment

The COVID-19 pandemic has completely changed remote work practices, shifting from only 15% of remote working hours pre-pandemic to over 50% of the American workforce working remotely by April 2020 [2]. The seismic shifts in the employment world have become very evident because remote work is taking up its proper role in the modern labor world. Most organizations have embraced flexible work arrangements by giving their employees the freedom to redefine their work-life balance through reduced commuting time and increased productivity in a comfortable environment. This is further evident in the trend to maintain home office spaces and flexible working hours which have become crucial for firms to sustain remote work practices in the new post-pandemic norm [14].

Moreover, the Low-Touch Economy has been a vital driver that has forced businesses to transform their models, leveraging a maximum amount of digital platforms but reducing the physical form of touching. Embracing such changes, therefore, will enable employers to efficiently find their way in the new turbulence of the labor market. Resilience and continued operation will be fostered as employers move on with their activities despite the changing economic platforms [22].

The pandemic not only disrupted some traditional employment practices but also ushered in a more flexible and technological workforce. This significant shift has uncovered several key trends in the labor market:

Adoption and Impact: Remote work adoption differs greatly among demographic and socioeconomic groups. White-collar employees and individuals with higher education have had better access to remote work opportunities, while service and manual labor roles encounter limited choices. Research suggests job losses were three times greater for non-remote workers in comparison to remote workers during the pandemic [40].

Organizational Response: Organizations have adjusted by creating hybrid work models that combine remote and in-office work. While some employees indicate increased productivity owing to shorter commute times, others find it difficult to manage challenges in their home environments. This has resulted in varied productivity outcomes and necessitated new management strategies [60].

However, this transition has also the potential to worsen socioeconomic disparities, as lower-income employees frequently hold positions that cannot be performed remotely. Achieving success in this evolving environment necessitates organizations to modify their management techniques and infrastructure while providing equitable access to opportunities across all segments of the workforce. This structural shift will have enduring consequences for urban development, transportation systems, and economic geography, requiring careful policy responses to guarantee inclusive economic growth in the post-pandemic period.

7.3.3 Working Poverty in Labor Markets

Working poverty, often referred to as "in-work poverty," is a multifaceted socioeconomic issue wherein individuals remain impoverished despite being employed. This concept gained prominence as a subject of academic inquiry in Europe during the late 1990s, challenging the conventional belief that having a job guarantees an escape from poverty [3]. As per The International Labor Organization, "the working poverty rate reveals the proportion of the employed population who are living in poverty despite being employed." Hence, underscoring the disconnection between participation in the labor market and overall economic security.

ILO COVID-19 Monitor indicates that the immediate repercussions of quarantine measures and reduced economic activity led to a loss of nearly 30,000 work months among the infected individuals, severely impacting the income security of workers. This disruption is underscored by estimated overall labor income losses ranging from 860 to 3440 billion USD. Contrary to the pre-pandemic projection of a decline in working poverty by 14 million, the crisis has instead resulted in an additional 8.8 million individuals entering working poverty, with severe scenarios potentially escalating this figure to between 20.1 and 35.0 million [59].

The structural factors contributing to working poverty disproportionately affect vulnerable populations, with significant variations across different demographics and income brackets. Young workers, who already contend with elevated rates of unemployment and underemployment, have become increasingly susceptible to diminished labor demand. Women have been particularly hard hit, with ILO data revealing that 58.6% of employed women are situated in the severely impacted services sector, in contrast to 45.4% of men. The intensity of these effects varies markedly among income groups, with lower-middle-income countries potentially witnessing increases of up to 14.8 million individuals in extreme and moderate working poverty under high-impact scenarios [3]. These trends indicate that effectively addressing working poverty necessitates comprehensive policy interventions that prioritize not only job creation but also the quality of employment, wage standards, and protections for workers.

7.3.4 Reflection on the Long-Term Implications for Employment Patterns and Labor Market Resilience

In the post-COVID-19 landscape, the long-term implications for employment patterns are evident. Work from Home and Hybrid work mode, which gained prominence during the pandemic, is likely to remain a key feature in many companies, offering both flexibility to the employee as well as the potential to reduce the cost incurred in office spaces and operations to firms.

Although this shift has its own set of challenges, especially in access and equity at work, as not everyone is readily afforded the correct technologically functional environments for working remotely [61]. In addition, sectors as primal as hospitality and travel, of course, directly damaged from the crisis, will find it harder and take longer to rehabilitate their frayed employment relationship, which may further augment job polarization [61]. While some jobs will change and new ones will come in related to technology and healthcare, the displaced individuals will face a huge reskilling curve to get back into the workforce. It is ultimately going to require education, training, and more importantly, all for excellent policies that support a diverse workforce to build resilience in labor markets.

7.3.4.1 Recovery Patterns and Policy Reforms

Diverse regions across the world adopted a variety of strategies to alleviate the labor market repercussions:

1. US and EU: Both regions instituted a multitude of initiatives aimed at providing support to employees and businesses. However, the labor market in the United States was impacted more acutely than that of the EU, which demonstrated a greater capacity to stabilize employment levels [36].
2. Russia: The labor market in Russia implemented measures such as wage reductions, part-time reallocations, and telecommuting options in an effort to avert an employment collapse and an escalation in unemployment, thereby reflecting a distinctive adaptive strategy [37].
3. Global Trends: The ILO emphasized the global nature of the crisis and highlighted the urgent need for continued tracking and analysis in order to develop appropriate policy responses to the situation [42].

As far as recovery and future projections are concerned, preliminary evidence from developing countries indicated a partial resurgence in employment rates between April and August 2020, with more substantial improvements observed among groups that initially endured the greatest job losses [41]. Nevertheless, the overarching labor market recovery remains ambiguous, with forecasts indicating persistent challenges and an imperative for sustained policy interventions [35].

7.4 Inflation Dynamics and Currency Market Volatility

From the last two decades or so, the inflation trend was low and stable in major advanced economies. However, it plummeted sharply in the COVID-19 Crisis. The recent surge in inflation caught many macroeconomists off guard, sparking an intense debate about how to explain this increase[24]. The economic shock caused by the pandemic brought a challenge, both for academics and policymakers, of describing and forecasting inflation rates. The years from 2020 to 2023 were marked by three different types of exogenous shocks: First being the production/consumption shock due to COVID-related lockdowns, another one is supply chain shock, and lastly an energy price shock because of the Russian invasion of Ukraine [39]. Altogether these three major shocks disrupted the economic equilibrium globally, causing inflationary pressures and exposing vulnerabilities in financial markets.

Inflation rates, the backbone of macroeconomic stability, deeply affect the purchasing power, the behavior of households toward different goods and services and investment decisions by firms. The recent trends of inflation peak and post Coronavirus spread have proposed macroeconomic models with a nonlinear Phillips curve that has a flat slope when the rate of inflation is low and under control and the curve steepens when inflationary pressures are higher than usual [24]. This nonlinearity arises due to a quasi-kinked demand schedule for goods produced by firms. These models suggest that all shocks in the economy transmit more strongly to inflation when inflation is surging, particularly cost-push shocks which can propagate even stronger when inflation is high and rising above its steady state level.

In parallel, the pandemic exposed the fragility of currency markets. Exchange rate volatility intensified as global trade and capital flows contracted sharply, reflecting the interconnected nature of modern economies. Heightened uncertainty led to abrupt currency depreciations in many emerging markets, compounding their economic challenges. Advanced economies also faced currency fluctuations, albeit to varying degrees, influenced by the scale and timing of fiscal and monetary interventions. For instance, while China's strong containment measures stabilized its currency, the US dollar experienced sustained volatility, highlighting the divergent impacts of COVID-19-related shocks on major economies [44]. Moreover, inflation and currency volatility are deeply intertwined, reflecting the broader macroeconomic environment. Rising energy prices during the pandemic, for example, had a dual effect: fueling inflationary pressures and increasing exchange rate fluctuations, particularly for energy-importing nations. This duality underscores the importance of managing inflation expectations and ensuring exchange rate stability as critical components of economic recovery.

The monetary policy response to this inflation surge has faced significant challenges. Research indicates that central banks face a more severe trade-off between inflation and output stabilization when inflation is elevated. This finding is driven by the fact that cost-push shocks propagate more strongly to inflation than monetary policy shocks when inflation is surging above the central bank's inflation target [24]. Without appropriate fiscal policy, the shocked economy can take years to recover, or even tip over into a deep recession. Moreover, the success of monetary policy in curbing inflation is primarily due to expectation anchoring, rather than the direct economic impact of interest rate hikes [39].

7.4.1 Inflation Trends Post Coronavirus Outbreak

In the immediate aftermath of the pandemic, there were sharp fluctuations in price levels. Recent research reveals that most of the price fluctuations was due to the supply-side factors rather than demand-pull inflation [52]. The disturbances in production and supply chains had a dominant role, as companies adjusted to different pricing strategies to cope up with these challenges.

As revealed in the dataset, at the peak of pandemic, world inflation rates were fairly stable at 3.3%. Developing economies experienced a higher inflation (5.2%) (Table 7.5). While inflation in advanced economies was quite low (0.7%). This stable inflation rate in 2020 can be attributed to low consumer demand and economic activity resulting from extensive lockdown measures.

The post-pandemic years marked a remarkable increase in inflation rates. World inflation rose 4.7% in 2021 and 8.6% in 2022. This surge was especially seen in advanced economies, from 3.1% in 2021 to 7.3% in 2022, a more than tenfold increase compared to earlier levels. The rates in emerging markets and developing economies elevated to 9.6% in 2022. This pattern reveals a notable transformation from the pandemic era to the subsequent post-pandemic years.

Table 7.5 Inflation rate, average consumer prices (Annual percent change)

Economy type	2020	2021	2022	2023	2024
World	3.3	4.7	8.6	6.7	5.8
Advanced economies	0.7	3.1	7.3	4.6	2.6
Emerging market and developing economies	5.2	5.8	9.6	8.1	7.9

Source: Data tabulated from IMF DataMapper [34]

The data for 2023 and 2024 implies subdued inflationary pressures as the world rates declined to 6.7% in 2023 and 5.8% in 2024. Developed nations saw a more pronounced decrease in inflation the year 2024, while the emerging markets and developing economies sustained higher inflation rates at 7.9%. This indicates that although the monetary policy actions have yielded results, the recovery pattern differs markedly between advanced and emerging countries.

7.4.2 Volatility, Spillover Effects, and Exchange Rate Regimes

Studies have identified significant volatility spillover effects in strategic commodity markets to have a significant impact on different inflation metrics of an economy. Rastogi and Kanoujiya [53] concluded that crude oil price volatility has long-run effects on inflation, especially concerning India. On the other hand, the volatility in gold price showed a bidirectional negative relationship with the wholesale price index. There has also been works on exchange rate volatility having a negative effect on inflation, specifically concerning the consumer price index of USD/INR [53]. Carriero et al. [11] suggested via their works that a substantial portion of inflation volatility can be attributed to global factors, like the

Producer Price Index (PPI) of China and oil bubble inflation. It was further identified that the inter-continent production networks' volatility had been a crucial determinant of both levels and persistence of inflation since the early 1990s, underscoring the increasing interconnectedness of the global economy. Differences in labor market institutions across countries, such as unemployment insurance schemes, can lead to varying inflation volatilities. Campolmi and Faia [9] argue that these institutional differences affect how real marginal costs and inflation react to area-wide shocks, resulting in welfare losses at the currency area level.

Several studies are also there to illustrate that the change in the exchange rate can also cause macroeconomic volatilities. The choice of the policy regime like domestic inflation targeting, Consumer Price Index (CPI) based Taylor rules, or exchange rate pegs result in different levels of exchange rate volatility. Galí and Monacelli [21] suggest domestic inflation targeting since it is often considered optimal for minimizing welfare losses associated with suboptimal rules. While Lim and Mcnelis [45] found that the counterfactual switch from a fixed exchange rate regime to domestic inflation targeting would reduce inflation slightly and increase the volatility of both types of interest rates significantly.

There are many studies which show that change in government spending and the monetary policy can, to a significant degree, affect macroeconomic volatility. Liu [46] reported the monetary policy to be the cause of a substantial portion of the output and inflation volatility explained by about 40% and 20%, respectively, while the impact of government spending, in addition to these volatilities, is considerable, and is firm less compared to the monetary policy shocks [46].

The macroeconomic volatilities, especially the consumption growth volatility, GDP growth volatility, and inflation volatility, can affect asset prices like the rate of return on equity, the interest rates on the bond or bill, or the exchange rate of a dollar with other currencies. Ermolov [16] shows that the equation system of the consideration of the macroeconomic volatility and the consumption asymptotic perfect foresight consumption model can better explain many puzzles in the stock market, bond market, and forex market. Also, Engle et al. [15] found macroeconomic fundamentals to have a large role in the stock market volatility even at short horizons.

7.4.3 Comparison of Inflation and Exchange Rates Before and After COVID-19 in Major Advanced and Emerging Economies

Table 7.6 Comparison of inflation and exchange rates pre and post COVID-19

Economy type	Period	Inflation dynamics	Exchange rate volatility
Advanced economies	Pre COVID-19	Relatively stable inflation rates, with low volatility in core G3 currencies (dollar, euro, yen) [27].	Exchange rate volatility was moving downward, reaching similar levels to the Bretton Woods era [27].
	Post COVID-19	Inflation remained mostly silent during the pandemic, with central banks focusing on economic activity [62]	Exchange rate volatility increased initially but was mitigated by government interventions and policies [18, 57].
Emerging economies	Pre COVID-19	Variable inflation rates compared to advanced economies, but generally stable [27].	Lower exchange rate volatility than historical averages, partially due to global monetary policy convergence [27].
	Post COVID-19	Inflation dynamics were less affected, but economic activity and currency depreciation were significant concerns [62].	Exchange rate volatility spiked due to increased economic uncertainty and varied government responses [18, 23].

Source: Made by Author based on referred research

7.5 Macroeconomic Policy Responses to COVID-19

The pandemic posed an extraordinary challenge for the policymakers and governments to make the right policy intervention choices in order to stabilize the corona hit economy. In macroeconomics there are two main pillars of policy intervention namely the fiscal policy monetary policy intervention. Fiscal policy includes government intervention for providing immediate relief in the forms of direct cash transfers, providing subsidies to households and firms and tax cuts. Fiscal intervention was heavily done in this crisis in various forms. Monetary policy, on the other hand, involves rate cuts, quantitative easing, and liquidity injections by central banks to stabilize financial markets and encourage lending. Moreover, exchange rate volatility increased initially but was mitigated by government interventions and policies as shown in Table 7.6.

Both approaches were crucial but operated under different constraints. Advanced economies had more fiscal space and developed financial systems, allowing for expansive stimulus packages and monetary easing. In contrast, emerging markets and low-income countries had to adopt more targeted measures due to limited fiscal resources and structural challenges. Despite these differences, the need for coordinated and comprehensive action was recognized globally to sustain economic activity and protect livelihoods.

This section is dedicated to delving deeper into these strategies adopted by various economies, highlighting their impact and limitations. By analyzing these responses, it will be better understood how nations have managed the challenges posed by the pandemic and these lessons might shape different policy frameworks in the future.

7.5.1 Fiscal Policy Responses

Globally, governments used fiscal policy tools to address and mitigate the initial economic impact of COVID-19. These included increased health expenditures, income transfers, welfare payments, and wage subsidies to minimize short-term unemployment [49]. The fiscal responses differed considerably across nations, with advanced economies (AEs) implementing more extensive measures compared to emerging market economies (EMEs) as they have a larger fiscal space and strong social safety net [1].

Several economies raised their direct spending and transfers to households and firms. For example, there were some policies that involved increasing government spending on goods and services by 2% of GDP and expanding infrastructure spending significantly [20]. In emerging economies, the goal was to provide emergency relief to vulnerable sections of their populations and affected businesses to avoid mass layoffs and bankruptcies [47]. Extensive fiscal stimulus packages were implemented, including direct spending, loans, credit facilities, tax cuts, and short-term loan extensions [55].

The fiscal indicators during peak COVID reveal some major differences in the fiscal capacity of different nations and accordingly their economic response for the crisis period (Table 7.7). A detailed analysis of these indicators is as follows:

Primary Net Lending/Borrowing Position: Fiscal deficit was the largest for the advanced economies as −8.94% of their annual GDP in 2020 followed by emerging economies as −6.91%, while low-income countries had smaller deficits, −3.5% of GDP. This pattern clearly reflects the varying fiscal space available to different nations. It is obvious that advanced economies have greater capacity to implement expansionary fiscal policies. In 2021 the data showed substantial improvement from −5.75%, −3.26%, to −2.64% respectively, suggesting a gradual fiscal consolidation across all economies.

Revenue and Expenditure Dynamics: A striking disparity exists in revenue generation capabilities. Advanced economies maintained robust revenue levels around 36% of GDP, while emerging markets generated about 26%, and low-income countries merely 13% of

Table 7.7 Major fiscal policy indicator for advanced, middle- and low-income economies peak COVID-19

	Advanced economies		Emerging market and middle-income economies		Low-income developing countries	
Indicator (as % of GDP)	2020	2021	2020	2021	2020	2021
Primary net lending/ borrowing	−8.94	−5.75	−6.91	−3.26	−3.5	−2.64
Revenue	36	36.87	25.49	26.56	12.84	13.45
Expenditure	46.18	43.96	34.19	31.59	18.15	18.07
Gross debt	121.84	115.4	65.53	64.72	49.11	49.11
Net debt	86.25	83.32	45.34	44.89	44.97	45.55

Source: extracted from IMF Datamapper, Fiscal Monitor Datasets [28]
Note: IMF Fiscal Monitor classifies the world into three major group economies which include 41 advanced economies, 96 emerging market and middle-income economies, and 58 low-income developing countries

GDP. This revenue gradient directly corresponds to institutional capacity and economic development levels.

The expenditure patterns too reveal similar disparities but also show the scale of fiscal response: In Advanced economies the expenditure was 46.18% of GDP in 2020 and 43.96% in 2021. The expenditure trend in emerging markets appeared as 34.19% of GDP in and somewhat lower in 2021 at 31.59%. However in low income economies the expenditure as percent of GDP remained at 18.15% in 2020 and 18.07% in 2021. The significant gap between revenues and expenditures, particularly in advanced economies, indicates the scale of fiscal stimulus deployed during the crisis.

Debt Burden Analysis: The debt burden patterns during COVID-19 reveal distinct fiscal capacities across economic groups. Advanced economies had the highest gross debt at 121.84% of GDP in 2020, which somewhat declined to 115.4% in 2021, showcasing their huge fiscal response and strong lending capacity. Developing nations maintained a moderate level at 65.53% of GDP in 2020, while low-income countries showed the lowest at 49.11%, indicating their borrowing constraint and restricted fiscal space.

The net debt positions provide more insights into financial health. Advanced economies' net debt at 86.25% of GDP in 2020 suggests substantial asset holdings offsetting their gross debt. The smaller gap between gross and net debt in low-income countries 49.11% versus 44.97% in 2020 indicates limited financial buffers. Emerging markets maintained a decent position with net debt at 45.34%, reflecting their moderate financial capacity and market access.

The fiscal response trend has several important implications for post-pandemic economic recovery. It reflects the need for international cooperation to assist and provide aid to lower-income countries in managing their fiscal targets. The varying levels of fiscal support also suggest potentially divergent recovery paths across different economic groups in the post-pandemic period.

7.5.2 Monetary Policy Responses

Central banks worldwide adopted expansionary monetary policies to support liquidity and stabilize financial markets. These measures included lowering policy rates, such as the repurchase agreement rate (repo) and cash reserve requirement (CRR), and offering credit facilities to meet the demand for loans and advances [55]. Monetary authorities financed rising public debt through asset purchases, a practice that was significantly boosted during the pandemic [13]. The rise of decentralized cryptocurrencies led some governments to consider creating centralized virtual currencies, known as CBDCs, to enhance monetary policy effectiveness [13].

7.5.2.1 Analysis of Monetary Policy Measures

The above table summarizing key measures offer a comparative look at how different economic groups responded to the crisis. This analysis discusses the broader and detailed policy response intervention based on the sources mentioned below the summary (Table 7.8).

Interest Rate Adjustments To reduce borrowing costs and kickstart economic activity, central banks extensively lowered down interest rates. Advanced economies cut their policy rates to almost zero, including the US, the Eurozone, and Japan for an extended period. The US Federal Reserve, for instance, changed its monetary policy stance from tightening to expansionary in March 2020 by lowering its target range to 0–0.25%.

For India, Brazil and other developing countries the task was to strike a balance between inflation control and economic stimulus along with higher pre-pandemic inflation rates. These economies kept a close eye on price stability even as they implemented sizable rate cuts.

To foster economic growth, the Reserve Bank of India, for instance, lowered its repo rate several times while keeping an eye out for potential inflationary pressures. In contrast, other emerging markets, like Brazil, reduced interest rates while keeping them higher than in developed nations to mitigate the risk of inflation spike and currency depreciation.

Lending Operations and Credit Facilities Central banks worldwide improved the lending facilities to provide easy flow of credit to households and firms. Many new initiatives were implemented by different nations as a policy response. The European Central Bank (ECB) introduced its Targeted Longer-Term Refinancing Operations (TLTROs) with an aim to give banks low-cost funding that was contingent on lending to the real economy. Likewise, to assist big businesses impacted by the pandemic, the Bank of England established the Covid Corporate Financing Facility (CCFF).

The Federal Reserve in the US also launched several initiatives, including the Main Street Lending Program for assistance of small and medium businesses. Targeted long-term repo operations (TLTROs) were implemented in emerging markets like India to facilitate liquidity into the areas that were most severely hit by the crisis, such as microfinance institutions and small enterprises. During the pandemic, these activities were essential for sustaining credit flow, averting a credit crunch, and boosting economic activity.

Table 7.8 Summarized key monetary policy measures during COVID-19 (March 2020–June 2021)

Measure	Advanced economies	Emerging markets and middle-income economies	Low-income developing economies
Interest rate cuts	Near-zero rates, US Federal Reserve cut to 0–0.25%	Moderate Rate Cuts to maintain inflation stability	Limited cuts due to Inflation concerns; constrained monetary space
Lending facilities	TLTROs by ECB, CCFF by Band of England	Liquidity facilities to support SMEs; TLTROs in India	Limited or absent
Reserve requirement adjustments	Rarely used due to developed financial markets	CRR cuts by RBI (100 bps; INR 1.37 trillion released)	Frequently used to boost liquidity
Foreign exchange interventions	Not common, except to stabilize investor confidence	Brazil and Indonesia intervened to curb currency Volatility	Used sparingly due to limited forex reserve
Asset purchase programs (QE)	Used Extensively; $4.6 trillion by Federal reserve	Targeted bond purchases to stabilize markets	Rarely used due to limited capacity and underdeveloped financial markets

Sources: IMF [29], European Central Bank [17] and Cant et al. [10]

Reserve Policies Maintaining a required reserve and adjusting it as per the need of the economy is an essential tool central banks have, especially in the emerging markets. In India for example, the Reserve Bank reduced the cash reserve ratio (CRR) by 100 basis points in 2020 to provide liquidity to the economy and banking system which released approximately INR 1.37 trillion into the Indian economy. This helped increase the availability of funds for lending purposes and to support various economic activities.

However, in advanced economies, reserve requirement adjustments were less common, as financial markets were already more liquid and flexible.

Reserve requirement reductions were effective in countries where banks were the primary channel for credit distribution. In emerging markets, this measure provided much-needed liquidity to the banking sector, supporting lending and economic stability. In contrast, advanced economies typically employed more sophisticated tools, making reserve adjustments less central to their responses.

Foreign Exchange and Market Stabilization Measures For the maintenance of exchange rate stability and to get and regain investors' confidence, several central banks intervene in foreign exchange markets. Emerging nations, such as Brazil and Indonesia, aggressively used their foreign exchange reserves to prevent currency depreciation, which causes inflation overall and economic instability. To mitigate global dollar shortage and to support the emerging market financial scenario, the Federal Reserve established swap lines with foreign central banks providing US dollar liquidity. Foreign exchange interventions were important for currency stability and prevention of capital outflows during the pandemic.

The effectiveness of these measures varied according to the type of economy. Advanced economies, with more significant foreign exchange reserves and better international financial markets access, faced less difficulties as compared to emerging and developing economies, where reserves were more limited, and the risk of currency depreciation was higher.

Asset Purchase Programs (Quantitative Easing) Quantitative Easing tools were used at a very large scale by the central banks to mitigate the crisis by stimulating liquidity in the economy. As are especially evident for the Advanced economies' monetary policy responses. The transmission of QE effects through financial markets revealed distinct patterns across economies. The European Central Bank's Pandemic Emergency Purchase Program (PEPP), totaled for €1.85 trillion, exemplified a more nuanced approach. According to the ECB Working Paper Series, while these interventions successfully stabilized sovereign debt markets, their impact on real economic activity varied significantly across sectors and regions. The ECB's experience particularly demonstrated how institutional frameworks influence QE effectiveness, with the monetary union's structure both enabling and constraining policy transmission.

The Bank for International Settlements' Annual Economic Report highlights striking contrasts in both scale and implementation between advanced and emerging market approaches to QE. The Bank of Japan's continued yield curve control alongside its QE program provided crucial insights about the limitations of monetary policy in a persistent low-interest-rate environment. Despite deploying additional pandemic-related purchases, Japan's modest economic recovery underscored the diminishing returns of prolonged monetary accommodation.

Emerging market central banks adopted more calibrated approaches, reflecting their unique constraints and vulnerabilities. The IMF's Global Financial Stability Report [32] documents how emerging markets implemented innovative adaptations of QE principles while managing the delicate balance between stimulus and currency stability.

7.5.2.2 Exit Strategies and Future Implications

According to the BIS Quarterly Review, the transition from emergency QE measures has revealed complex challenges in policy normalization. These experiences underscore how exit strategies must carefully consider:

1. Market Stability: Preventing disorderly market reactions while unwinding positions.
2. Inflation Management: Balancing price stability with growth objectives.
3. Financial System Resilience: Maintaining system stability during policy normalization.
4. International Spillovers: Managing cross-border impacts of policy shifts.

7.5.3 Lessons Learned and Conclusions

The World has never been the same again after the COVID-19 pandemic. This disease has caused tremendous damage to people, their quality of life, and physical and emotional well-being. For policymakers and economists, it has presented a very challenging situation due to its severe and long-term economic consequences that directly impact day to day activities of people and society. The present chapter has tried to provide an in-depth analysis of the major macroeconomic factors that were disrupted by the pandemic. From its impact on growth and GDP dynamics, labor market shocks, inflation, to macroeconomic policy responses, categorizing economies into advanced, emerging, and low-income/high-income groups to capture the diverse experiences and outcomes. The analysis demonstrated that while advanced economies were better positioned to implement large-scale fiscal and monetary interventions. Because of their limited financial and fiscal resources, low-income developing economies have implemented fewer and less comprehensive policies, but they have nevertheless increased liquidity through the use of instruments like reserve requirement adjustments. Together with fiscal responses, these concerted efforts played a key role in reducing the severity of the crisis and laying the groundwork for future economic growth.

Across all economic groups, the COVID-19 pandemic prompted a wide range of monetary policy actions. In order to successfully stabilize their financial systems and aid in recovery, advanced economies made use of their institutional capacity to execute extensive quantitative easing (QE) programs, interest rate reductions, and lending facilities. However, the long-term effectiveness of these measures varies, with many emerging and low-income economies still grappling with the economic scars of the pandemic. Inflationary pressures, workforce disruptions, and growing income inequality are pressing challenges that policymakers must address as part of their recovery efforts. While implementing comparable strategies, emerging markets faced more difficulties managing external vulnerabilities and inflation. The aggressive fiscal responses have led to highly elevated public debt levels, posing future macroeconomic risks. There is a consensus that fiscal consolidation, rather than further fiscal stimulus, will be necessary to address the fiscal legacy of the crisis [49]. The challenge lies in balancing fiscal adjustment with maintaining economic growth momentum while avoiding the exacerbation of existing inequalities.

The need for international cooperation is emphasized, particularly for EMEs constrained by high public debt and external financial limitations. Coordinated efforts led by the G20 and supported by the International Monetary Fund (IMF) are crucial to enable these countries to undertake necessary fiscal responses [50]. This becomes especially critical as advanced economies' policy decisions create potential spillover effects on emerging markets. In many developing countries, the effectiveness of monetary and fiscal stimulus is limited due to weak monetary transmission mechanisms and smaller fiscal multipliers. Therefore, the focus should be on avoiding procyclicality, ensuring the continuity of public services, and supporting vulnerable populations [47]. The experience highlights the importance of building robust institutional frameworks capable of responding to future shocks while maintaining economic stability.

References

1. Alberola E, et al. Fiscal response to the COVID-19 crisis in advanced and emerging market economies. Pac Econ Rev. 2021;26(4):459–68.
2. Angelucci M, et al. Remote work and the heterogeneous impact of COVID-19 on employment and health. National Bureau of Economic Research; 2020. Available at: http://www.nber.org/papers/w27749
3. Arnania-Kepuladze T, Sologashvili D, Kepuladze G. FBIM transactions labor market: COVID-19 pandemic and working poverty. 2021. Available at: https://www.researchgate.net/publication/366866413_FBIM_Transactions_LABOR_MARKET_COVID-19_PANDEMIC_AND_WORKING_POVERTY. Accessed 29 Nov 2024.
4. Bell DNF, Blanchflower DG. Us and UK labour markets before and during the Covid-19 crash. Natl Inst Econ Rev. 2020;252:R52–69.
5. Bluedorn J, et al. Gender and employment in the COVID-19 recession: evidence on "She-cessions". IMF Working Pap. 2021;2021(095):1.
6. Borio C. The Covid-19 economic crisis: dangerously unique. Bus Econ (Cleveland, Ohio). 2020;55(4):181–90.
7. Bruinen de Bruin Y, et al. Initial impacts of global risk mitigation measures taken during the combatting of the COVID-19 pandemic. Saf Sci. 2020;128:104773.
8. Byrialsen MR, Olesen F, Madsen MO. The macroeconomic effects of covid-19: the imperative need for a Keynesian solution. Rev Régul. 2021;29(29). Available at: https://journals.openedition.org/regulation/18405
9. Campolmi A, Faia E. Labor market institutions and inflation volatility in the euro area. J Econ Dyn Control. 2011;35(5):793–812.
10. Cant C, et al. A global database on central banks' monetary responses to Covid-19. Monetary and Economic Department; 2021. Available at: https://elischolar.library.yale.edu/ypfs-documents/13385. Accessed 30 Nov 2024.
11. Carriero A, Corsello F, Marcellino M. The global component of inflation volatility. SSRN Electron J. 2018; Available at: https://doi.org/10.2139/ssrn.3176957.
12. Chudik A, et al. A counterfactual economic analysis of COVID-19 using a threshold augmented multi-country model. SSRN Electron J. 2020; Available at: https://doi.org/10.2139/ssrn.3695670.
13. Echarte Fernández MÁ, et al. Central banks' monetary policy in the face of the COVID-19 economic crisis: monetary stimulus and the emergence of CBDCs. Sustainability. 2021;13(8):4242.
14. EKOMBIS Review. Jurnal Ilmiah Ekonomi dan Bisnis. Unived Press; 2024. Available at: https://core.ac.uk/download/616686213.pdf
15. Engle RF, Ghysels E, Sohn B. Stock market volatility and macroeconomic fundamentals. Rev Econ Stat. 2013;95(3):776–97.
16. Ermolov A. Macroeconomic volatility and asset prices. Columbia University; 2015. Available at: https://doi.org/10.7916/D88G8JTR.
17. European Central Bank. Pandemic emergency purchase programme. European Central Bank; 2024. Available at: https://www.ecb.europa.eu/mopo/implement/pepp/html/index.en.html. Accessed 30 Nov 2024.
18. Feng G-F, et al. What is the exchange rate volatility response to COVID-19 and government interventions? Econ Anal Policy. 2021;69:705–19.
19. Fernandes N. Economic effects of Coronavirus outbreak (COVID-19) on the world economy. SSRN Electron J. 2020; Available at: https://doi.org/10.2139/ssrn.3557504.
20. Fernando R, McKibbin WJ. Macroeconomic policy adjustments due to COVID-19: scenarios to 2025 with a focus on Asia. SSRN Electron J. 2021; Available at: https://doi.org/10.2139/ssrn.3774611.
21. Galí J, Monacelli T. Monetary policy and exchange rate volatility in a small open economy. SSRN Electron J. 2004; Available at: https://doi.org/10.2139/ssrn.859564.
22. Gigauri I. New economic concepts shaping Business Models in post-pandemic era. Int J Innov Technol Econ. 2021;1(33) Available at: https://doi.org/10.31435/rsglobal_ijite/30032021/7393.
23. Gunay S. Comparing COVID-19 with the GFC: a shockwave analysis of currency markets. Res Int Bus Financ. 2021;56:101377.
24. Harding M, Lindé J, Trabandt M. Understanding post-COVID inflation dynamics. J Monet Econ. 2023; Available at: https://doi.org/10.1016/j.jmoneco.2023.05.012.
25. Huang Y. Comments on the global macroeconomic impacts of COVID-19: seven scenarios. Asian Econ Pap. 2021;20(2):38–40.
26. ILOSTAT Data Explorer. 2024. Available at: https://rshiny.ilo.org/dataexplorer36/?lang=en&segment=indicator&id=UNE_2EAP_SEX_AGE_RT_A
27. Ilzetzki E, Reinhart CM, Rogoff KS. Will the secular decline in exchange rate and inflation volatility survive COVID-19? Brook Pap Econ Act. 2020;2020(3):279–332.
28. IMF. Fiscal policies database. International Monetary Fund; 2020a. Available at: https://

www.imf.org/en/Topics/imf-and-covid19/Fiscal-Policies-Database-in-Response-to-COVID-19
29. IMF. Policy responses to COVID19. International Monetary Fund; 2020b. Available at: https://www.imf.org/en/Topics/imf-and-covid19/Policy-Responses-to-COVID-19
30. IMF. World economic outlook database, October 2020. International Monetary Fund; 2020c. Available at: https://www.imf.org/en/Publications/WEO/Issues/2020/09/30/world-economic-outlook-october-2020
31. IMF. World economic outlook database, October 2021: recovery during a pandemic. International Monetary Fund; 2021. Available at: https://www.imf.org/en/Publications/WEO/Issues/2021/10/12/world-economic-outlook-october-2021
32. IMF. World economic outlook database, October 2022. International Monetary Fund; 2022. Available at: https://www.imf.org/en/Publications/WEO/weo-database/2022/October
33. IMF. World economic outlook database, October 2023. International Monetary Fund; 2023. Available at: https://www.imf.org/en/Publications/WEO/Issues/2023/10/10/world-economic-outlook-october-2023
34. IMF. International Monetary Fund (IMF). 2024. Available at: https://www.imf.org/en/Home
35. International Labour Organization. World employment and social outlook: trends 2022. Geneva: ILO; 2022. Available at: https://doi.org/10.54394/dspl5113.
36. Jolović N. Labor market in conditions of COVID-19 pandemic: EU vs. USA. Ekonom Teor praksa. 2020;13(3):13–33.
37. Kapeliushnikov RI. The anatomy of the corona crisis through the lens of the labor market adjustment. Vopr Ekon. 2022;2:33–68.
38. Khan A, Khan N, Shafiq M. The economic impact of COVID-19 from a global perspective. Contemp Econ. 2021;15(1):64–75.
39. Knicker MS, et al. Post-COVID inflation and the monetary policy dilemma: an agent-based scenario analysis. J Econ Interac Coord. 2024; Available at: https://link.springer.com/article/10.1007/s11403-024-00413-3.
40. Kong X, et al. Work from home in the post-COVID world. Case Stud Transp Policy. 2022;10(2):1118–31.
41. Kugler M, et al. How did the COVID-19 crisis affect different types of workers in the developing world? World Dev. 2023;170(106331):106331.
42. Lee S, Schmidt-Klau D, Verick S. The labour market impacts of the COVID-19: a global perspective. Indian J Labour Econ. 2020;63(Suppl 1):11–5.
43. Lenzen M, et al. Global socio-economic losses and environmental gains from the coronavirus pandemic. PLoS One. 2020;15(7):e0235654.
44. Li C, et al. COVID-19 and currency market: a comparative analysis of exchange rate movement in China and USA during pandemic. Econ Res. 2022;35(1):2477–92.
45. Lim GC, Mcnelis PD. Macroeconomic volatility and counterfactual inflation-targeting in Hong Kong: macroeconomic adjustment in Hong Kong. Pac Econ Rev. 2012;17(2):304–25.
46. Liu X. On fiscal and monetary policy-induced macroeconomic volatility dynamics. J Econ Dyn Control. 2021;127(104123):104123.
47. Loayza NV, Pennings S. Macroeconomic policy in the time of COVID-19: a primer for developing countries. Washington, DC: World Bank; 2020.
48. Ludvigson S, Ma S, Ng S. COVID-19 and the macroeconomic effects of costly disasters. Cambridge, MA: National Bureau of Economic Research; 2020. Available at: https://doi.org/10.3386/w26987.
49. Makin AJ, Layton A. The global fiscal response to COVID-19: risks and repercussions. Econ Anal Policy. 2021;69:340–9.
50. McKibbin WJ, Vines D. Global macroeconomic cooperation in response to the COVID-19 pandemic: a roadmap for the G20 and the IMF. SSRN Electron J. 2020; Available at: https://doi.org/10.2139/ssrn.3651586.
51. Msemburi W, et al. The WHO estimates of excess mortality associated with the COVID-19 pandemic. Nature. 2023;613(7942):130–7.
52. Orszag PR, Brooks R, Murdock WE III. COVID-19 inflation was a supply shock. Brookings. 2024. Available at: https://www.brookings.edu/articles/covid-19-inflation-was-a-supply-shock/. Accessed 30 Nov 2024.
53. Rastogi S, Kanoujiya J. The volatility spillover effect of macroeconomic indicators and strategic commodities on inflation: evidence from India. S Asian J Bus Stud. 2024;13(2):180–200.
54. Rungcharoenkitkul P. Macroeconomic effects of COVID-19: a mid-term review. Pac Econ Rev. 2021;26(4):439–58.
55. Sarker PK. COVID crisis: fiscal, monetary and macro-financial policy responses. SSRN Electron J. 2020; Available at: https://doi.org/10.2139/ssrn.3601524.
56. Sone E, Ko J-H. COVID-19 pandemic and its economic impact on the global economy: a computable general equilibrium approach. Coronaviruses. 2024;5(2) Available at: https://doi.org/10.2174/012666797526969323 1107072320.
57. Umar Z, Gubareva M. A time-frequency analysis of the impact of the Covid-19 induced panic on the volatility of currency and cryptocurrency markets. J Behav Exp Financ. 2020;28(100404):100404.

58. UN Statistics Division. The Sustainable Development Goals report 2022: goal 1—end poverty in all its forms everywhere. 2022. Available at: https://unstats.un.org/sdgs/report/2022/goal-01/#:~:text=This%20erased%20more%20than%20four,to%20the%20pre%2DCOVID%20trend
59. Valensisi G. COVID-19 and global poverty: are LDCs being left behind? Eur J Dev Res. 2020;32(5):1535–57.
60. Vyas L. "New normal" at work in a post-COVID world: work–life balance and labor markets. Polic Soc. 2022;41(1):155–67.
61. World Economic Forum. The future of jobs report 2020. 2020. Available at: https://www.weforum.org/publications/the-future-of-jobs-report-2020/in-full/chapter-1-the-labour-market-outlook-in-the-pandemic-economy/
62. Yilmazkuday H. COVID-19 and monetary policy with zero bounds: a cross-country investigation. Financ Res Lett. 2022;44(102103):102103.

The Pandemic Crisis and Its Implications for Fiscal Policies

8

Poorna Shri, Samrah Rehan, Saurabh Mishra, Vishal Chand, Ashish Srivastava, Ankita Raghav, and Supriya Raghav

8.1 Introduction

8.1.1 An Overview of the COVID-19 Pandemic as a World Shock That Has Never Been Seen Before

Health systems, economies, and communities are all being impacted at the same time and on a never-before-seen scale by the COVID-19 pandemic, which is brought on by the SARS-CoV-2 virus. In late 2019, the virus was discovered in Wuhan, China. It quickly spread to other continents, and in March 2020, the World Health Organization (WHO) declared it to be a worldwide pandemic [92, 93]. Table 8.1 depicts the series of events occurred in India during the COVID-19 pandemic. Unlike previous financial, geopolitical, or environmental crises, COVID-19 occurred without warning, giving little time for preparation, and it impacted practically every nation, regardless of its level of wealth or development. This crisis was particularly distinctive because it was multifaceted—as it sparked a

Table 8.1 Chronological overview of major COVID-19-related events and government responses in India from January 2020 through mid-2021

Date	Event/Response
30 Jan 2020	First confirmed COVID-19 case in Kerala, India
11 Mar 2020	WHO declares COVID-19 a global pandemic
24 Mar 2020	India imposes nationwide lockdown
Apr–May 2020	Migration crisis due to lockdown
Jun 2020	Unlock phases begin
Early 2021	Second wave hits India with Delta variant
Jan 2021	Launch of COVID-19 vaccination drive
May–June 2021	Peak of second wave, severe healthcare burden

Authors "Poorna Shri", "Samrah Rehan" and "Saurabh Mishra" have equally contributed to this chapter.

P. Shri (✉) · S. Rehan
Department of Biochemistry, GSVM Medical College, Kanpur, Uttar Pradesh, India

S. Mishra · V. Chand
School of Life Sciences and Biotechnology, CSJM University, Kanpur, Uttar Pradesh, India
e-mail: vishalchand@csjmu.ac.in

A. Srivastava
Department of Zoology, Siddarth University, Siddharth Nagar, Uttar Pradesh, India
e-mail: ashishsri@suksn.edu.in

A. Raghav
Deprtment of Economics, IIMT University, Meerut, Uttar Pradesh, India

S. Raghav
Department of Public Health, Poornima University, Jaipur, Rajasthan, India

A. Raghav et al. (eds.), *COVID-19 Impact on Economy, Environment and Healthcare*,
https://doi.org/10.1007/978-981-95-1196-9_8

humanitarian disaster, a global supply and demand shock, a public health emergency, and a financial depression all at once. Officially, there were over 6.9 million fatalities in 2023; however, estimates of excess mortality indicate that number could be significantly higher [94]. The largest collapse in peacetime in over a century occurred in 2020, when the global GDP shrank by about 3.1% [39–42]. In 2020, 255 million full-time jobs were lost, increasing fourfold the effects of the 2008–2009 financial crisis, according to the International Labour Organization [36, 37].

To stop the virus from spreading, lockdowns and other measures were put in place, which resulted in the closure of businesses, the closure of borders, and the suspension of international trade and travel. Entire industries collapsed overnight, including the travel, aviation, and hotel sectors. Food insecurity and inflation were made worse in many regions of the world by shortages in the supply chain that developed as manufacturing closed and transportation lagged [2, 3]. The private sector, which employs more than 60% of the global workforce, was particularly negatively impacted due to the lack of social safeguards and digital infrastructure [78–82]. Additionally, COVID-19 revealed and made existing imbalances worse. Access to immunizations, healthcare facilities, and financial resources was more challenging for low-income nations. Within nations, migrant labor, women, and ethnic minorities were among those with limited opportunities that experienced disproportionate health and economic consequences [57–62]. Wealth imbalances increased but at the same time technology-driven industries grew like stock markets recovered swiftly, and the use of digital platforms increased.

More importantly, the pandemic revealed the weaknesses in the existing structures for global governance. The early battle for medical supplies, vaccine nationalism, and fragmented policy coordination underscored the need for a more resilient and cooperative global health and economic system [25]. In conclusion, COVID-19 was a worldwide shock comparable to the Spanish flu of 1918 or the World Wars but with significantly more interlinked effects because of globalization. Fiscal, social, and public health policies were reconsidered as a result, and the need for flexible, inclusive, and long-lasting policy solutions has been brought to spotlight.

8.1.2 The Significance of Comprehending Fiscal Policy Reactions During Economic and Public Health Emergencies

It is crucial to fully understand how fiscal policy reacts to economic and public health emergencies to build strong economic systems, minimize instability in society, and prepare for the next crisis. Together, taxation and spending by the government form fiscal policy, which can be used to increase demand, save lives, and support essential services during emergencies. When supply and demand were affected simultaneously, as was the case during the COVID-19 pandemic, the significance of fiscal actions became even more critical [32]. The primary purpose of fiscal policy is to maintain economic stability in the face of unfavorable macroeconomic circumstances. Governments can counterbalance decreasing private sector activity and prevent economic collapse by improving income and expanding public spending. This is particularly crucial during public health emergencies when healthcare demands and lockdowns lower economic productivity and consumption [26]. Early fiscal action can reduce the intensity and length of recessions and avoid long-term economic damage.

Understanding the type, scope, and timing of fiscal interventions can also help governments avoid mistakes in the past and identify best practices. For instance, fiscal policies that are poorly targeted or delayed may cause wasteful spending, delay recovery, or raise inequality. Knowing how successful specific fiscal measures, like cash transfers, wage subsidies, or investments in healthcare facilities, are can help governments modify their policies according on the type of crisis and institutional environment [27]. Social cohesiveness and public trust—two things that are crucial in times of crisis—are supported by efficient fiscal policy. According to the IMF Fiscal Monitor [39–42], quick support to people with limited resources reduces the risk of civil disturbance, enhances compliance with public health measures, and bolsters the legitimacy of political institutions.

Additionally, analyzing financial responses during crises provides important insights into sustainability and budgetary space. Many governments had to balance mounting public debt with urgent spending demands, and while emergency investment is often justified, long-term sustainability requires careful planning of exit strategies, tax reforms, and expenditure reprioritization [57–62]. Lessons learned from similar crises can inform the development of automatic stabilizers, or systems that respond without legislative delays, such as unemployment insurance or food aid. Lastly, cross-national comparisons and data-driven assessments of fiscal actions contribute to the development of global economic resilience. Coordinated fiscal actions can support recovery efforts and lessen cross-border spillovers, especially in interconnected economies. For instance, developing nations who lacked autonomous fiscal capacity were mostly dependent on multilateral aid from organizations such as the World Bank and IMF during COVID-19 [84–89]. In conclusion, crisis management, as well as recovery and preparedness following a disaster, depend on knowing and evaluating fiscal policy actions during times of crisis. It provides the analytical tools and statistical information that decision-makers require to ensure future fiscal responses are more successful, inclusive, and timely.

8.2 Analysis of the Function, Efficacy, and Prospects of Fiscal Policies Following the Pandemic Is the Chapter's Main Goal

The above approach enables governments to better understand how resilient fiscal frameworks may help them plan for and respond to future global shocks, whether they are connected to the economy, the environment, or public health.

(a) *Function of Fiscal Policies*

The role that fiscal measures played during the epidemic was crucial in mitigating its effects on the economy and society. Emergency measures like wage subsidies, cash transfers, tax deferrals, and higher public health spending were immediately implemented by governments. In order to maintain businesses, protect household income, and strengthen health systems, these actions were implemented in response to an uncommon equivalent supply-demand shock [2, 3]. Understanding these goals makes it easier to evaluate how well fiscal tools fulfilled their said functions and how much they acted as immediate stabilizers.

(b) *The Effectiveness of Fiscal Measures*

Considering short-term economic stabilization results including GDP recovery, employment trends, poverty alleviation, and business continuity, it is necessary to evaluate the success of these fiscal measures. For example, Europe's furlough policies helped reduce unemployment surges, while the United States' CARES Act greatly stabilized income levels during lockdowns [11, 12]. However, institutional capacity, fiscal headroom, and the nature of informal sectors significantly affected how effective fiscal measures were, especially in low- and middle-income countries [39–42]. These measures are compared in this chapter along with their primary benefits and drawbacks.

(c) *Design of Future Fiscal Policy*

Because of the epidemic, fiscal ideology has changed, leading to a reassessment of traditional guidelines, debt tolerance, and the role of governments in maintaining macroeconomic stability. Following a pandemic, fiscal policies are anticipated to support a number of objectives, including inequality reduction, digital transformation, climate resilience, and economic recovery. To ensure medium-term fiscal sustainability, deploy automatic stabilizers during crises, and align expenditure with the Sustainable Development Goals (SDGs), it is necessary to develop flexible fiscal frameworks, according to Blanchard and Pisani-Ferry [10].

The objectives of these policies were

- Analyze how different nations' fiscal responses to COVID-19 were structured.
- Analyze their effects on inequality, fiscal sustainability, and socioeconomic stability, now and in the future.
- Make recommendations for policies that will lead to strong, equitable, and sustainable fiscal systems.
- Consider how the changed global climate could influence fiscal policy.

8.3 The Pandemic's Economic Repercussions

8.3.1 Worldwide Supply Networks and Trade Disruptions, a Global GDP Downturn, and an Increase in Unemployment

A record-breaking number of global supply chains and international trade were disrupted by the COVID-19 pandemic, which led to a rapid increase in unemployment and a coordinated decline in global GDP. The severity and complexity of the economic consequences revealed important fault lines in the world's labor and production markets and highlighted the vulnerabilities of an extremely interconnected global economy.

(a) *Disruptions Affecting the World Economy and Supply Chain*

Significant disruptions were experienced by global supply chains, especially those that relied on timely production and international outsourcing. The early lockdowns in industrial hubs such as China, followed by a series of restrictions in other countries, resulted in shortages of intermediate products, shipping challenges, and production delays in several industries [55]. Supply chains in the electronics, pharmaceutical, and automobile sectors were particularly affected due to their dependence on foreign inputs.

Restrictions on travel and the export of vaccines and medical supplies increased the stress on international collaboration and brought attention to the need for robust distinct supply chains. In 2020, worldwide trade volumes dropped by 5.3%, one of the biggest drops since the 2008 financial crisis, according to the World Trade Organization [95].

(b) *Contraction in Global GDP*

Both developed and developing nations saw a sharp decline in economic output as production stopped and demand fell. Global GDP fell by about 3.1% in 2020, according to the International Monetary Fund [39–42]. Advanced economies including the United States, the European Union (EU), and Japan experienced significant declines in output, while emerging markets and developing economies (EMDEs) also faced challenges like unpredictable commodity prices, reduced remittance payments, and capital outflows.

The pandemic-induced recession was distinct because it was caused by a demand shock from lockdowns and a decline in consumer confidence, as well as a supply shock from production halts and labor force reductions [4]. Long-term effects were seen in employment and investment in the service sectors, especially in the travel, hospitality, and aviation sectors.

(c) *Unemployment and Labor Market Shocks*

The labor markets around the world were seriously impacted. According to the International Labour Organization, the loss of 255 million full-time jobs in 2020 was four times as severe as the global financial crisis in 2009 [36, 37]. Young workers, women, and those in informal employment were disproportionately affected. Because they lacked social protection and job security, workers in the informal sector—which accounts for over 60% of the global workforce—were especially susceptible. While furloughs and pay subsidies in high-income

Table 8.2 Multidimensional socioeconomic consequences of the COVID-19 pandemic in India, highlighting the vulnerable sectors most affected

Sector	Key impacts
Employment	Job losses, especially in informal sector; increase in unemployment
Education	School closures, shift to online learning, widening of the digital divide
Health	Overburdened healthcare system; decline in routine and non-COVID services
Economy	GDP contraction, disrupted supply chains, business shutdowns
Migration	Mass migration of laborers to rural areas amid lockdown
Gender	Rise in domestic violence; increased unpaid caregiving burden on women

countries helped mitigate the effects, many low- and middle-income countries had the financial means to maintain stable labor markets, leading to long-term negative consequences such as increased poverty and inequality [78–82] (Table 8.2).

8.4 Effects on Sectors (e.g., Manufacturing, Hotels, Healthcare)

Industries that primarily rely on global supply chains and interpersonal relationships were particularly vulnerable to the COVID-19 pandemic's varied consequences on various economic sectors. While certain sectors, like technology and healthcare, experienced unheard-of expansion and change, others witnessed devastating losses. Important insights into the wider economic impact and recovery paths can be gained by looking at these sector-specific effects.

(a) *Manufacturing Industry*

The industrial sector saw major disruptions due to workforce shortages, plant closures, and initial supply chain problems. Early in the pandemic, industry output declined sharply worldwide, with sectors including electronics, automotive, and aerospace being especially hard-hit because of their reliance on international supply chains [13]. In response, firms adopted automation, digital technology, and localized manufacturing methods to increase the robustness of their supply chains. According to UNIDO [71], although manufacturing returned in late 2020, the recovery was uneven, with emerging economies seeing slower recoveries due to finance and infrastructure constraints.

(b) *Tourism and Hospitality Sector*

Among the industries most adversely impacted were hospitality, tourism, and travel. In 2020, the United Nations World Tourism Organization reports that foreign tourist visits decreased 74% from the previous year [83]. Lockdowns, travel restrictions, and consumer concern led to massive revenue losses, several insolvencies, and the loss of millions of jobs in the food, hospitality, and aviation sectors. The sector has recovered slowly and has been centered on changing consumer choices, travel laws, and immunization rates. A lot of companies in this industry changed their focus to contactless services, digital interaction, and health safety standards in an effort to win back customers.

(c) *Healthcare Industry*

There was a remarkable rise in the demand for resources and services in the healthcare industry. Medical personnel, ventilators, intensive care unit (ICU) beds, and personal protective equipment (PPE) were all in limited supply in hospitals across the globe [92, 93]. Concurrently delaying or stopping non-COVID medical services, like cancer screenings, elective operations, and mental health therapies, raised long-term public health issues.

Positive aspect of the pandemic inspired innovation in healthcare delivery, especially in the areas of vaccine development, telemedicine, and remote monitoring [49]. The quick creation and distribution of COVID-19 vaccinations demonstrated the effectiveness of international scientific cooperation and public-private collaborations.

(d) *E-commerce and Technology Sectors*

During the epidemic, technology and e-commerce flourished as many other sectors collapsed. Internet-based video conferencing, digital payment systems, and online retail companies experienced immense growth as businesses and consumers adapted to digital living and remote work [52]. The long-term effects of this digital acceleration are expected to include higher investments in cyber security, artificial intelligence, and e-learning systems.

8.5 The USA, EU, and India Are a Few Examples of Big Economies

Nation's economic size, fiscal space, and institutional capabilities all had an impact on the scope and structure of their fiscal reactions to the COVID-19 an outbreak. Among the major economies, some of the most important and varied fiscal interventions were carried out by the US, the EU, and India.

(a) *United States*

To minimize the economic impact of the pandemic, the US government implemented a number of extreme fiscal measures between 2020 and 2021, costing more than $5 trillion [15–17].

Among the important programs were:

- The $2.2 trillion Coronavirus Aid, Relief, and Economic Security (CARES) Act (March 2020) provided direct cash transfers, expanded unemployment insurance, improved healthcare providers, and provided business loans.
- A further $1.9 trillion was provided to the economy by the American Rescue Plan Act (March 2021), which gave priority to direct payments, vaccine distribution, child tax credits, and support for state and local governments.
- These policies accelerated economic recovery, prevented widespread unemployment, and stabilized consumption, but they also increased the national debt.

(b) *The European Union*

The European Union launched a coordinated budgetary response to support member states, based on the historic Next Generation EU recovery plan at €750 billion [21–23]. Among the highlights are:

- National recovery plans are supported by grants and loans to member states that emphasize digital and green transformations.
- Nations' capacity to run higher deficits due to flexibility in the application of the Stability and Growth Pact's fiscal constraints.
- Each of the member nations has increased spending on healthcare, salary subsidies, and business support programs, like as Germany's €130 billion stimulus plan and Italy's "Cura Italia" edict.
- Within the EU framework, the collective approach was a major step toward stronger budgetary unity.

(c) *India*

India had two challenges: managing a large informal sector and limited fiscal headroom. A package of approximately ₹29.87 trillion, or nearly 15% of GDP, was used to launch the Atmanirbhar Bharat Abhiyan, also referred to as the "Self-Reliant India Campaign" [53, 54] (Table 8.3).

Key components included:

Table 8.3 Major policy measures and financial relief packages implemented by the Indian government to mitigate the impact of the pandemic on health and livelihoods

Initiative	Description
Pradhan Mantri Garib Kalyan yojana	Free ration and cash transfers for poor and migrant households
Atmanirbhar Bharat Abhiyan	Economic revival stimulus with credit support for MSMEs
PM CARES fund	Special emergency relief fund for pandemic response
Vaccination drive	Phased national program for COVID-19 vaccine distribution

- The Pradhan Mantri Garib Kalyan Yojana provides food distribution and direct financial transfers to underprivileged communities.
- Initiatives to improve agriculture and labor regulations, as well as lending assistance to small and medium-sized enterprises (SMEs).
- Expenditures by the government on healthcare facilities and initiatives to boost manufacturing, such as Production Linked Incentives (PLI).
- Longer-term structural adjustments intended to encourage economic stability were combined with short-term stimulus in India's approach.

8.6 The Role of Fiscal Policy in Emergencies

8.6.1 Fiscal Policy and Its Goals

Fiscal policy is how the government uses taxes and spending to influence a nation's economic activity. According to Blanchard and Johnson [9], it is a key component of macroeconomic management that is utilized to accomplish wealth redistribution, promote growth, and stabilize economic cycles. Unlike monetary policy, which is often under the control of central banks, fiscal policy is guided by governments and parliaments, making it a highly charged instrument.

(a) *Primary Objectives of Fiscal Policy*

Fiscal policy is employed to reduce instability during times of economic recession or overheating. The purpose of expansionary fiscal measures, like higher government spending or tax breaks, is to stimulate demand during recessions. On the other hand, restrictive fiscal measures assist in cooling the economy during times of inflation or overheating [39–42].

(b) *Distribution of Resources*

Fiscal policy may increase productivity and long-term economic growth by concentrating resources on critical areas like innovation, infrastructure, health, and education. Public investment often stimulates private sector development [56].

(c) *Distribution of Incomes*

Social harmony and economic inequality can be improved by fiscal policy through the implementation of welfare programs, social transfers, and progressive taxation. In rising nations and during times of economic hardship, this is particularly crucial [57–62].

(d) *Full-Time Employment*

Maintaining high employment rates is one of fiscal policy's main socioeconomic objectives. Corporate support and public works initiatives are crucial for job creation and recovery after disruptions like the COVID-19 outbreak [50, 78–82].

(e) *Sustainability of Public Debt*

Additionally, fiscal policy seeks to maintain long-term, reasonable levels of governmental debt. Fiscal expansions that are poorly managed, especially in emerging nations, can result in unmanageable deficits and higher borrowing costs [90, 91].

Types

1. Expansionary fiscal policy involves reducing taxes and/or increasing governmental spending in order to increase demand during recessions.
2. Contractionary fiscal policy is used to regulate the economy by increasing taxes and/or reducing public spending when inflation is excessive.
3. Neutral fiscal policy refers to preserving a balance between government revenue and spending, typically in the context of a stable economy.

8.7 During the Epidemic, Distinguish Between the Responsibilities of Fiscal and Monetary Policy

With the COVID-19 pandemic, two unique difficulties emerged: a health crisis and a significant economic punch. Important but distinct roles were played by monetary and fiscal policies in this well-coordinated macroeconomic policy response. Understanding the distinct roles that governments and central banks played during the crisis, often working together to preserve economic stability and protect livelihoods, is made easier.

8.7.1 The Direct Economic Role of Government in Fiscal Policy

National governments employ fiscal policy through borrowing, taxation, and public expenditure to influence economic results. During the epidemic, fiscal policy emerged as the primary tool since it could be used to immediately address the crisis's effects on the real economy.

The Fundamental Responsibilities of Fiscal Policy during the Pandemic:

1. Provide Income support and employment protection and for that Governments provided direct transfers, salary subsidies, and unemployment insurance to households and workers. To support vulnerable people, for instance, the Pradhan Mantri Garib Kalyan Yojana (PMGKY) in India and the U.S. CARES Act offered significant financial infusions [39–42].
2. Investing in public health because the major component of budgetary initiatives was the emergency funding for vaccinations, personal protective equipment (PPE), and hospital capacity. Spending on health-related initiatives contributed to a decrease in viral transmission and death [57–62].
3. Assistance for Businesses and Sectors so that Loan guarantees, grants, and tax concessions were put in place to assist SMEs and important sectors such as tourism and aviation. The UK's vacation program and Germany's Kurzarbeit both helped to preserve professional relationships [36, 37].
4. Long-Term Recovery Initiatives: Fiscal policy enabled investments in infrastructure and green transition projects. The Next Generation European fund, for example, has a budget of €750 billion [21–23].

8.8 Central Bank's Function in Maintaining Liquidity and Stability in Monetary Policy

Central banks regulate monetary policy, which tries to regulate inflation, interest rates, and the money supply. Despite having less direct resources, monetary policy was extremely helpful throughout the pandemic.

Important Monetary Policy Roles during the Pandemic:

8.8.1 Maintaining Liquidity

Through emergency credit facilities, discount window lending, and open market operations, central banks gave banks and other financial institutions access to liquidity. The U.S. Federal Reserve extended its repo operations and launched a number of loan facilities [24].

8.8.2 Interest Rate Reduction

Many central banks cut policy rates to all-time lows in an effort to lower borrowing costs and boost demand. The European Central Bank (ECB) raised quantitative easing and kept interest rates negative, for instance [19].

8.8.3 Asset Acquisitions and Market Consistency

Financial markets were kept operational through asset acquisition and quantitative easing initiatives. As a result, governments were able to borrow money at a low cost to finance fiscal stimulus and credit shortages were avoided [6].

8.8.4 Management of Inflation and Exchange Rates

In the early phases of the pandemic, central banks also took action to prevent economic crises and control unstable exchange rates.

8.8.5 Synchronization and Integrity

Monetary and fiscal policy collaborated during the pandemic despite having distinct mandates. Fiscal stimulus performed better when financial conditions were stable and interest rates were kept low by accommodating monetary policy. Because of this cooperation, the crisis was prevented from getting worse and becoming a long-lasting depression. However, pressures have been exacerbated by rising inflation and interest rates since the outbreak, leading central banks to tighten policy while governments continue to face pressure to continue providing fiscal support for vulnerable populations [43–45].

8.9 Direct Transfers, Salary Subsidies, Tax Deferrals, and Emergency Assistance Are Examples of the Evolution of Fiscal Measures

A remarkable global economic response was triggered by the COVID-19 pandemic as governments tried to preserve livelihoods, save lives, and stabilize economies. Populations and economies were impacted by lockdowns, supply chain disruptions, and economic contractions, and fiscal policies swiftly evolved to satisfy those demands. The four primary fiscal intervention categories that dominated the policy environment were direct cash transfers, wage subsidies, tax deferrals, and emergency assistance programs. These were all modified to address specific pandemic-related concerns.

1. *Direct Cash Transfer:* The direct transfer of monies was one of the simplest and most visible budgeting measures. Governments provided households, particularly low-income and vulnerable groups, with direct cash aid in an effort to stabilize consumption and prevent poverty from worsening. Direct transfers prevented further declines in demand and served as a buffer for household expenditure during the peak of mobility restrictions [29].
   - Direct benefit transfers (DBTs) were made available to 200 million women and pensioners in India through the Pradhan Mantri Garib Kalyan Yojana (PMGKY) [31].
   - According to the Congressional Budget Office [14], the CARES Act (2020) in the United States provided stimulus checks totaling $1200 for adults and $500 for children.
   - According to the World Bank [84–89], the Auxílio Emergencial program in Brazil provided monthly compensation to millions of informal laborers.
2. *Wage Subsidies:* The implementation of wage subsidy programs by numerous governments aimed to prevent massive job cuts by rewarding companies to retain employees during periods of reduced business activity.

   These subsidies benefited in maintaining job connections during lockdowns to prevent extended unemployment and encourage a quicker recovery [57–62].
   - According to HM Treasury [34], the UK's Coronavirus Job Retention Scheme (furlough program) paid up to 80% of workers' wages.
   - In Germany, the traditional Kurzarbeit (short-time labor) program was expanded to include partial pay compensation [35].
   - The JobKeeper Payment was introduced in Australia to provide businesses with a subsidy to pay their employees [1].
3. *Tax Deferrals:* In order to reduce the immediate financial strain, governments offered tax deferrals and payment relief after realizing that businesses were seriously short on cash. Tax deferrals provided businesses with important short-term liquidity support, allowing them to stay afloat during periods of income

loss without having to declare bankruptcy or fire staff [39–42].

- Payroll tax deferrals and extended filing deadlines were offered [47], in the United States.
- Companies were allowed to defer paying corporate income taxes and social security payments in Italy and France [21–23].
- The Canada Emergency Rent Subsidy and tax benefits for small businesses were introduced in Canada [30].

4. *Emergency Assistance Programs:* Programs for emergency assistance rapidly expanded to accommodate individuals and groups who are not normally protected by social safety nets, such as unorganized laborers and freelance operators. Emergency assistance programs improved social safety nets during critical phases of the crisis, delivering essentials and temporarily lifting millions out of poverty [78–82].
   - South Africa was the first country to award the COVID-19 Social Relief of Distress Grant [84–89].
   - Through Japan's Special Cash Payment Program, all citizens received a universal cash transfer [48].

8.10 A Comparative Study of Pandemic-Induced Fiscal Measures

I. *A Comparative Analysis of Fiscal Responses in Various Nations*

Due to inequalities in institutional strength, public health priorities, and economic competence, the fiscal reactions of various nations to the COVID-19 pandemic differed significantly. The fact that wealthy nations frequently adopted more radical and expansive fiscal policies than did developing and low-income countries highlights the global differences in crisis management resources.

India

The Atmanirbhar Bharat package was one of India's responses; it aimed to offer free food distribution to underprivileged groups, loan guarantees for small businesses, and liquidity support, however with additional restrictions because of financial limitations. The fiscal measures were backed by extensive monetary and policy support, despite the fact that the direct fiscal spending was relatively low (between 1% and 2% of GDP) [53, 54].

United States

With a combined worth of more than $5 trillion, the American Rescue Plan (2021) and the CARES Act (2020) are two of the nation's most important fiscal stimulus initiatives ever. The proposals included direct cash transfers, small business loans (PPP), higher unemployment insurance, and funding for vaccines [15–17].

European Union

To prioritize investments in healthcare, digital transformation, and green recovery, the EU established the Next Generation EU, a €750 billion collective recovery fund. Member states also created their own national assistance plans; Germany, for instance, introduced fiscal stimulus equivalent to 10% of its GDP [21–23].

Japan

According to the IMF Fiscal Monitor [39–42], when loans and guarantees are taken into account, Japan experienced one of the biggest fiscal reactions in relation to GDP, estimated at almost 40%. The country implemented a number of stimulus plans worth over $3 trillion, with a focus on household subsidies, corporate support, and healthcare system stabilization.

Developing Economies

South Africa, Brazil, and Indonesia also implemented emergency relief measures, but with much smaller fiscal packages (2–5% of GDP), as a result of swelling debt loads and shrinking revenue bases. In 2020, however, Brazil's Auxílio Emergencial program temporarily moved millions out of poverty [84–89].

Key points

- Advanced nations received significantly more support (8–10% of GDP) than emerging markets (2–4% of GDP) [39–42].

- The richest countries mostly relied on direct transfers and health grants, lower-income nations relied more on financial assistance and liquidity measures.
- Countries like Estonia and South Korea that already had digital infrastructure had faster benefit distribution.
- While emergency spending was required, it greatly increased the national debt, raising concerns about the economy's long-term sustainability.

II. *Comparing Developed and Rising Economies*

In terms of fiscal capacity, institutional resilience, and economic recovery paths, the COVID-19 pandemic demonstrated and highlighted the significant differences between rich and emerging economies.

Advanced economies:

- Developed economies, like the United States, the European Union, Japan, and Canada, quickly and comprehensively implemented extensive fiscal assistance. In 2020–2021, these nations' average budgetary response, excluding additional guarantees and contingent liabilities, was around 8–10% of GDP [39–42].
- Direct subsidies, unemployment insurance, and cash transfers were prioritized in order to preserve household incomes and business viabilities are Large-Scale Direct Support.
- Healthcare Investments like expensive amounts are spent on vaccine procurement, testing facilities, and healthcare infrastructure.
- Focus on economic recovery in order to promote long-term growth; stimulus packages included investments in green energy, digital transformation, and infrastructure upgrades [63–65].
- By buying assets and cutting interest rates, central banks continued to adopt accommodative policies to assist fiscal initiatives to maintain Monetary-Fiscal Coordination.

Developing and Emerging Markets: South Africa, Indonesia, India, Brazil, and other low-income countries faced major challenges as emerging markets.

1. Smaller fiscal packages, which made up only 2–4% of GDP on average, were the result of limited fiscal flexibility and pre-existing high debt levels [84–89].
2. The primary goals of the programs were to guarantee food security, offer employment guarantees (such as India's MGNREGA), and provide liquidity support to small businesses.
3. Many of them were financed by external sources, including debt suspension programs (like the G20 DSSI program), IMF emergency financial facilities, and foreign aid.
4. It was challenging to implement assistance measures successfully due to the size of informal economies, inadequate healthcare systems, and a lack of administrative skills [74].

Comparative Perspectives:

1. *Fiscal Room:* When the pandemic arrived, industrialized countries were able to offer more stimulus without immediately experiencing solvency problems because they had lower interest rates and greater borrowing capacity [66].
2. *Digital Advantage:* Nations with strong digital payment systems, like the US, EU, and South Korea, may be able to provide aid more rapidly and efficiently.
3. *Recovery Speed:* The rate of recovery is higher in industrialized nations than emerging markets because to vaccination inequities and capital flight.
4. *Financial Risks:* Emerging economies are especially susceptible to financial hardship, as more than 60% of low-income countries are either in or at high risk of suffering it [33, 43–45].

III. *Packages of Stimuli (Scale, Timing, Structure)*

The massive financial instability brought on by the COVID-19 epidemic prompted stimulus programs to be implemented by governments of all shapes, sizes, and configurations. These stimulus programs were essential for preserving public health, lessening the immediate effects on the economy, and laying the foundation for future growth.

Different countries have quite different stimulus packages, primarily because of variations in fiscal capability, economic structure, and political will.

- The United States implemented a number of relief initiatives, including the CARES Act and the American Rescue Plan, which collectively contributed around 25% of GDP [15–17].
- Apart from national-level programs, the European Union created the "Next Generation EU" recovery fund, which has a value of €750 billion.
- Responses in emerging economies were lower in relation to GDP. Although the Atmanirbhar Bharat stimulus program received almost 10% of India's GDP, the majority of its funding came from credit guarantees rather than actual spending [39–42].
- The efficacy of fiscal measures was significantly influenced by timing-.
- As lockdowns began in early 2020, a number of countries quickly enacted emergency spending on healthcare, unemployment insurance, and direct transfers.
- Sustained assistance (2020–2021): In light of the crisis's prolonged duration, more rounds of budgetary assistance were introduced. For instance, more relief measures were authorized by the US in late 2020 and early 2021.
- Slow Reactions: The effectiveness of stimulus initiatives was diminished in a number of developing countries due to administrative and financial obstacles that delayed their implementation [84–89].

The fiscal packages were designed with both immediate needs and long-term goals in mind:

- Many nations provided direct cash transfers (such as the United States' Economic Impact Payments and Brazil's "Auxílio Emergencial"), boosted unemployment payments, and food help to households.
- In order to keep businesses solvent, business support programs included grants, loans, and salary subsidies (like the UK's Coronavirus Job Retention Scheme).
- A substantial amount of funds was allocated to secure vaccines, support public health programs, and strengthen healthcare systems.
- A number of countries provided specific support to sectors such as aviation, tourism, and hospitality.
- Some packages came with funding for renewable energy, digitalization, and infrastructure to help ensure a strong and long-lasting recovery [63–65].

The Function of Global Financial Organizations (World Bank, IMF)

Global financial institutions like the World Bank and the International Monetary Fund (IMF) were crucial in helping low- and middle-income countries, in particular, deal with the economic and health issues during the COVID-19 epidemic.

1. *The World Bank:* During the pandemic, the World Bank prioritized assisting public health systems, safeguarding the weak, and reviving economies.

   Emergency Health Finance: In order to support over 100 countries, the World Bank authorized a $160 billion COVID-19 response package that includes technical assistance, policy recommendations, and fast-track emergency finance.

Vaccine Access: The World Bank assisted in funding the acquisition and distribution of vaccines in developing nations through programs such as COVAX and its Multilateral Leaders Taskforce.

Economic Recovery and Social Protection: Money was utilized to help small enterprises and job development, bolster social safety nets, and give cash transfers.

2. *The International Monetary Fund (IMF):* During the outbreak, the IMF's primary function was to help nations in stabilizing their economy by offering financial support, policy recommendations, and technical assistance.

   Rapid Credit Facility (RCF): Through the Rapid Credit Facility (RCF) and Rapid Financing Instrument (RFI), the IMF provided emergency liquidity, allowing nations to get low-conditionality, fast credit. The IMF had given 85 nations more than $105 billion in emergency aid by the middle of 2021 [39–42].

   Debt Relief: The IMF expanded its Catastrophe Containment and Relief Trust (CCRT) to provide debt payment relief to assist the poorest countries in reallocating finances to pandemic response programs [39–42].

   Special Drawing Rights (SDRs): In August 2021, the IMF distributed an unprecedented $650 billion in SDRs to boost global liquidity and support recovery efforts, particularly in countries with low foreign reserves [39–42].

   Both groups collaborated with other international organizations (such as the World Health Organization, Gavi, and regional development banks) to guarantee a coordinated worldwide reaction.

   In order to free up funds for pandemic-related expenses, the World Bank and IMF led the G20 Debt Service Suspension Initiative (DSSI), which temporarily halted debt repayments for low-income nations [84–89].

Important points:

- Helped weaker economies avoid more severe financial crises.
- Made vital resources for immunization and medical treatment available.
- Facilitated the slow recovery in many developing countries and contributed to the preservation of macroeconomic stability.
- However, some complaints about:
- The aid's sufficiency in relation to the crises' magnitude.
- The need for more long-term, low-interest loans, or concessional financing.
- Fears that wealthier nation over-benefited from the SDR allotment.

8.11 Principal Obstacles in the Execution of Fiscal Policy Throughout the Epidemic

I. *Sustainability of the National Debt and Fiscal Room*

   Sustainability: The ability of a government to meet its debt service commitments (principal and interest payments) over an extended period of time without significantly increasing the risk of default or necessitating significant policy changes that could impede economic growth is known as sustainability of national debt. If a country can fulfill its present and future payment obligations without going into default or requiring extraordinary financial help, its debt is said to be sustainable. The sustainability of the national debt is influenced by several factors:

   1. *Economic Growth:* In relation to the size of the economy (as indicated by GDP), higher economic growth improves a country's capacity to generate income, which facilitates debt management and

reduction. One basic guideline that is frequently mentioned is that in order for debt to be sustainable, the rate of economic growth should ideally be greater than the debt's interest rate (g > r).

2. *Fiscal Policy:* The sustainability of debt depends on prudent fiscal policies, such as controlling budget deficits and maintaining a steady income base. Sustainability becomes more difficult when persistently high deficits result in rising debt levels [8].
3. *Interest Rates:* The cost of paying off debt is directly impacted by interest rates. Higher interest rates increase the cost of debt servicing, which puts a pressure on public resources.
4. *Debt Structure:* The type of debt, especially its currency denomination (domestic versus foreign) and maturity structure (short-term versus long-term), may have an effect on sustainability. A country may be vulnerable to exchange rate swings and refinances if a large portion of its debt is short-term or foreign currency [18].
5. *Market Confidence:* Investors' trust in a government's ability to manage its debt is a crucial component. A drop in confidence might make debt more unsustainable by raising borrowing costs and causing capital flight.

   Fiscal Room: Also known as fiscal space is the capacity of a government to enact new spending programs or reduce taxes without jeopardizing its financial stability or the stability of the economy. It represents the budgetary "wiggle room" that a government has to deal with unanticipated events (such natural disasters or economic crises) or to pursue capricious policy objectives. A government's fiscal room is determined by a number of factors, including:

6. *Current Fiscal Situation:* In general, a government has more fiscal flexibility when it has a stable revenue base, controllable deficits, and low debt levels.
7. *Future Fiscal Projections:* The estimation of fiscal room is influenced by projections of future revenue trends, spending requirements, and economic growth.
8. *Debt Sustainability:* A nation's financial flexibility will be constrained if its debt is excessive or headed in an unsustainable direction.
9. *Market Access and Funding Costs:* A government that has easy access to financial markets and cheap borrowing costs is usually more able to borrow money or make new expenditures.
10. *Regulatory Factors:* By promoting market confidence, sound fiscal institutions, open budget procedures, and a believable dedication to fiscal restraint can increase fiscal space.

II. *Effective Allocation of Financial Assistance*

Effective allocation is essential to preventing severe economic damage, lowering inequality, and reestablishing better institutions during emergencies like the COVID-19 epidemic. Financial aid must be deliberately allocated to the sectors, regions, and populations that require it most in order to maximize economic recovery, stabilize vulnerable groups, and promote long-term resilience. This is referred to as "effective allocation of financial assistance." Important guidelines for efficient allocation consist of:

Accountability and Transparency: Ensuring that resources reach their intended recipients and reducing corruption need transparent allocation.

Duration and Speed: In order to avoid job losses, company closures, and societal discontent, money must be deployed promptly.

Identifying the most prone to vulnerability: Small enterprises, low-income households, unorganized laborers, and healthcare systems with limited resources should be given priority when it comes to financial aid.

Variability: Real-time data regarding the changing nature of the crisis should be used to inform program adaptations.

A sustainable future: Support is to promote long-term structural changes in addition to offering short-term relief (e.g., investment in healthcare, digital infrastructure, and education).

According to Gentilini et al. [28], nations that instituted wage subsidy programs (such as Germany's "Kurzarbeit") or targeted cash transfers (such as Brazil's "Auxílio Emergencial") were able to mitigate the effects of the epidemic more successfully than those that used less focused interventions.

Additionally, the World Bank and the International Monetary Fund (IMF) stressed that delayed or poorly targeted fiscal support could hinder recovery and worsen inequality [39–42, 46].

III. *Disparities in Access and Inequality in Aid Initiatives*

Disparities in access to financial aid and public assistance programs frequently become glaring during international crises like the COVID-19 epidemic, revealing and escalating pre-existing social, economic, and geographical inequities.

Digital Divergence: A lot of aid initiatives, such as online cash transfer registrations, were run on digital platforms, which excluded people without internet access or digital literacy, especially in underserved and rural areas [76, 77].

Informal Market Elimination: Unemployment benefits and wage subsidies frequently ignored workers in the informal sector who did not have official social security numbers or registration [36, 37].

Regional Disparities: Regional imbalances grew as urban centers generally received aid more quickly and with stronger resources than rural or distant areas.

Gender Inequalities: Women encountered more barriers to receiving financial assistance and recovery programs since they are overrepresented in the unorganized and caregiving sectors [84–89].

Administrative Constraints and Corruption: In many nations, elites were able to seize help and misallocate resources due to weak administrative frameworks, a lack of transparency, and corruption, which disadvantaged people who needed it the most [69].

Few examples are:

- Many migrant workers in India were initially excluded from the Pradhan Mantri Garib Kalyan Yojana because they lacked official documentation, despite the fact that the program offered emergency cash transfers [70].
- Since more than 80% of workers in sub-Saharan Africa are employed in the informal sector, very few people there have access to institutional fiscal support programs [84–89].

8.12 Evaluation of Impact and Effectiveness of Policy

I. *Results in the Short Term: Poverty Reduction, Sectoral Revitalization, and Recovery Indicators*

At the onset of the COVID-19 pandemic, countries implemented large-scale budget stimulus initiatives. Significant but erratic short-term results were obtained from these efforts, particularly in the areas of poverty reduction, sectoral revitalization, and early recovery indices.

1. *Reduction of Poverty:* Unemployment benefits, food assistance programs, emergency cash transfers, and salary subsidies prevented millions of people from falling into poverty right away.
   - The World Bank [84–89] reports that although worldwide extreme poverty increased for the first time in 20 years in 2020, fiscal measures stopped an even bigger increase, preventing an estimated 20–30 million people from sliding below the international poverty line.
   - Programs such as the U.S. CARES Act and Brazil's Auxílio Emergencial

greatly reduced income losses for the lowest-income groups [29].

2. *Development of Sectors:* Depending on specific fiscal policies, different industries displayed differing levels of recovery.
   - *Healthcare:* Significant investments increased hospital capacity, diagnostics, and immunization programs, which momentarily increased employment in the healthcare sector [63–65].
   - *Retail and Hospitality:* Payroll support and small business grants helped many tourism, hospitality, and retail enterprises in nations including the United States and Australia weather the initial lockdowns [39–42].
   - *Industrial Sectors:* The recovery of industrial sectors was accelerated by stimulus aimed at supply chain resilience, such as support for semiconductors in East Asia.
3. *Indicators of Recovery:* After fiscal actions, important short-term economic indices showed signs of stabilization.
   - *GDP Growth:* The United States and Germany, two nations with substantial stimulus programs, recovered to their pre-pandemic GDP levels more quickly than other nations [43].
   - *Employment Rates:* During the height of the crisis, initiatives such as pay subsidies (such as Germany's Kurzarbeit) helped maintain unemployment below projections [63–65].
   - *Consumer Confidence:* By late 2020, consumer confidence indices in a number of economies had recovered, thanks to fiscal support that helped sustain consumer demand [68].
   - *Inflation Pressures:* The quick recovery raised spending, but it also triggered early signs of inflation, particularly in the US and emerging markets [7].

II. *Macroeconomic Modeling and Empirical Case Studies:*

Macroeconomic modeling and empirical case studies have been essential in assessing how well fiscal solutions worked during the COVID-19 pandemic. These instruments offer valuable perspectives on the causal effects of policy interventions on important economic metrics like inflation, growth, employment, and poverty.

1. *Macroeconomic Modeling:* Macroeconomic models have been used to forecast the fiscal effects and recovery scenarios of the pandemic, including structural VAR models, dynamic stochastic general equilibrium (DSGE) models, and computed general equilibrium (CGE) models.
   - The IMF [38] simulated the impact of fiscal stimuli using its G20MOD and came to the conclusion that prompt, extensive fiscal actions encouraged a faster recovery in output across G20 economies and avoided a deeper recession.
   - Using the QUEST model, the European Commission [20] predicted that, if well executed, the EU's Next Generation EU fund could increase the EU's GDP by up to 1.5% yearly by 2024.
   - Macro-simulation exercises created by the World Bank in 2021 demonstrated that extreme poverty, especially in low-income nations, would have increased much more sharply in the absence of significant fiscal actions.
2. *Empirical Case Studies:* To assess fiscal interventions, a number of case studies pertaining to particular nations have been carried out.
   - *United States:* Direct cash transfers under the CARES Act considerably enhanced short-term consumer spending, particularly among low-income households, according to research by Chetty et al. [11, 12], which evaluated the impact of fiscal stimulus using real-time economic tracker data.
   - *India:* According to the World Bank [84–89], vulnerable households were protected by the Pradhan Mantri Garib Kalyan Yojana (PMGKY), which comprised direct cash transfers and free

food grains. However, coverage gaps persisted, particularly in rural regions.

- *Germany:* Balleer et al. [5] showed that the country's Kurzarbeit wage subsidy program successfully reduced job losses, bringing unemployment down by roughly 2–3 percentage points as compared to a scenario in which no intervention was implemented.
- *Japan:* According to an empirical assessment by Kubota et al. [51], unconditional cash transfers of 100,000 yen per resident reduced household savings behavior while increasing household consumption.

8.13 Suggestions for Fiscal Policy

I. *Creating Inclusive and Robust Fiscal Systems*

Serious weaknesses in fiscal systems around the world were shown by the COVID-19 epidemic, underscoring the pressing need to develop inclusive and strong fiscal frameworks. Future fiscal policies should focus on boosting resilience to shocks, ensuring that benefits are widely and equitably dispersed across communities, and stabilizing the economy.

The Value of Inclusion in Financial Structures: Fiscal policy that is inclusive makes ensuring that the benefits of economic expansion and recovery are distributed fairly among various social groups, especially marginalized and vulnerable groups. A fiscal system that is inclusive:

- Increases access to social safety nets such as food security, unemployment insurance, and healthcare [28].
- Focuses funding on rural infrastructure, healthcare, and education in order to lessen long-term disparities [90, 91].
- Uses progressive taxation to pay for public services and goods without putting an undue burden on the poor [39–42].

The epidemic showed that nations with more robust social safety nets—like the Western nations—were better equipped to safeguard their citizens from economic shocks.

Worldwide Outlook

Following the pandemic, both the UN and the IMF advocated for a "build forward better" approach that prioritized inclusivity and resilience over merely restoring pre-crisis conditions [39–42, 72, 73]. Leading countries, such as Germany, with its commitment to green recovery investment, and New Zealand, with its "Wellbeing Budget," provide practical examples of efforts to integrate these concepts into fiscal policymaking.

II. *Incorporating Digital Infrastructure to Implement Policies Quickly and Transparently*

The necessity of digital infrastructure in allowing governments to react to public health and economic crises in a timely, effective, and transparent manner was highlighted by the COVID-19 pandemic. Nations with strong digital infrastructures were better equipped to provide their citizens with information, healthcare, and financial assistance, reducing the negative effects of the crisis.

The Function of Digital Infrastructure in Crisis Management: Digital infrastructure, which includes open data portals, cloud-based public finance management, digital payment platforms, and electronic identity systems (e-ID), is essential for:

- Governments were able to minimize delays by promptly reaching beneficiaries using digital cash transfer systems. For example, during the pandemic, millions of people were able to receive direct benefit payments because to India's Aadhaar-enabled payment system [84–89].
- Digital platforms that track fund transfers and public procurement in real-time minimize corruption and misuse potential [78–82].
- As seen in nations like Kenya (via M-PESA), mobile money and digital wallets have made financial services more accessible in underserved and distant areas, promoting inclusive service delivery [67].

Essential Elements of a Successful Digital Infrastructure

- Governments may precisely target people for benefits while preventing duplication by using unique digital IDs.
- For smooth and inclusive payment delivery, government databases, mobile networks, and banking are integrated.
- Governments may better manage resources and anticipate requirements with the use of predictive analytics.
- As our dependence on digital systems increases, it is more crucial than ever to protect data security and privacy [57–62].

Although digitization has many advantages, there are drawbacks as well:

Digital Divide: Disparities in digital literacy and internet access can make marginalization of vulnerable populations worse.

Privacy Issues: To protect individual rights, extensive personal data collecting necessitates stringent measures.

Institutional Readiness: To properly utilize digital tools, governments need to make investments in technical infrastructure, change management, and human capacity [39–42].

III. *Coordinating Fiscal Policy with the SDGs and Other Long-Term Development Goals*

The essential necessity for governments to coordinate fiscal policy with both short-term economic stabilization goals and long-term sustainable development objectives, especially the United Nations Sustainable Development Goals (SDGs), was emphasized by the COVID-19 pandemic. In order to achieve inclusive, sustainable growth and solve global issues like poverty, inequality, climate change, and health crises, fiscal policy—a key instrument for resource mobilization and allocation—is essential.

1. *The Significance of SDG Alignment*

   Fiscal policy that is in line with the SDGs guarantees that efforts to recover from crises will help create a more resilient, just, and ecologically sustainable society. Sustainable debt management techniques, progressive taxation, and targeted public spending can all be beneficial.

   - Eliminate poverty (SDG 1).
   - Assure high-quality education and healthcare (SDGs 3 and 4).
   - Encourage gender parity (SDG 5).
   - Encourage economic growth and decent work (SDG 8).
   - Encourage sustainable industry and construct robust infrastructure (SDG 9).
   - Combat global warming (SDG 13) [72, 73].
2. *Fiscal-SDG Coordination Techniques*
   - Green budgeting is the practice of nations like France and Ireland that integrate environmental sustainability assessments into their budgetary procedures [63–65].
   - Increasing funding for social safety, healthcare, and education is a top priority for social spending and is essential to reaching the SDGs [43–45].
   - The goal of progressive tax reforms is to create tax structures that increase resources without making inequality worse [78–82].
   - Improving efficiency and directing public investment toward areas that directly support the accomplishment of the SDGs is known as public investment management.
3. Identification and Responsibility: It is essential to build strong mechanisms for impact analysis and budget tracking. Measures like SDG Budget Tagging, which Mexico and Nepal have implemented, aid in tracking the allocation of public funds to SDG-related initiatives [76, 77].
4. *Implementation Difficulties:*
   - *Limited Fiscal Space:* Many developing nations are unable to increase SDG-oriented investments due to their high debt levels.
   - *Policy Coherence:* It's still difficult to create synergies and steer clear of trade-offs between various policy goals (such as growth vs. environmental protection).

- *Data and Measurement Gaps:* Insufficient timely and disaggregated data makes it difficult to effectively evaluate budgetary contributions to the SDGs.

5. *Worldwide Instances:*
   - *European Union:* In line with the SDGs, the Next Generation EU recovery plan places a strong emphasis on digital and green transitions.
   - *Rwanda:* Set an example for policy consistency by integrating the SDG targets with its national budget framework [75].
   - *Bhutan:* Keeps its fiscal policies in line with the Gross National Happiness philosophy, striking a balance between social and environmental well-being and economic progress.
6. *Policy Recommendations:*
   - A clear connection between specific SDG targets and tax and spending policies should be made in fiscal strategy.
   - Promote private sector investments that align with national SDG priorities.
   - Encourage international cooperation by providing debt relief, technical support, and concessional financing to low- and middle-income nations in order to free up funds for SDG initiatives.

8.14 Conclusion

Fiscal policy is crucial to crisis management, as the epidemic served as an unpleasant reminder. The need for a more inclusive global fiscal architecture is highlighted by disparities in fiscal competence, despite the fact that many measures were successful in averting a catastrophe. The lessons learnt from this age can be used to inform future policy, which balances immediate alleviation with long-term stability and sustainability. Future plans should focus on improving economic diversity, improving social safety nets, and reorganizing global supply chains in order to increase tolerance to future shocks. Global economic interdependence was shown to be fragile during the COVID-19 pandemic, which also caused supply chains to break, trade to collapse, global output to fall, and unemployment to rise.

The worldwide outbreak changed the dynamics of industries around the world. Industries that relied on face-to-face interactions saw longer downturns, whereas technology-driven sectors experienced more rapid growth. Recovery plans must therefore be sector-specific, emphasizing digital change, durability, and equitable support for the most affected industries. Despite the implementation of considerable budgetary adjustments in all three economies, the design and priority differed depending on internal issues and policy frameworks. The EU combined solidarity with strategic investment in future growth, India mixed short-term relief with longer-term economic improvements, and the United States gave direct fiscal stimulus to people and businesses first priority. These varied experiences offer valuable advice on how to modify fiscal policy in response to systemic shocks. In order to achieve macroeconomic stability, equitable development, and sustainable growth, governments employ fiscal policy as a crucial instrument. In times of disaster like the COVID-19 pandemic, it is especially important for protecting lives and livelihoods. Nonetheless, the effectiveness of fiscal policy is influenced by the government's institutional capacity, transparency, and prompt implementation. Although monetary and fiscal policy both have been equally important in lessening the economic effects of COVID-19, they both served essentially different purposes. Monetary policy ensured the stability and liquidity of the financial system, but fiscal policy had a direct role in providing relief and stimulating demand. An organized effort ensured a quick and effective response, highlighting the importance of specific policy duties in future worldwide recessions. Economic policy development during the COVID-19 pandemic demonstrated extraordinary innovation and diversity. Instead of using typical stimulus strategies, governments used more targeted, direct measures to protect homes, preserve jobs, and support businesses. Despite variations in their breadth and pace from country to country, these measures served to lessen the worst economic and social repercussions of the pandemic. By learning from these experiences, budgetary structures that are more crisis-resistant can be developed.

Having sufficient fiscal space is based upon the sustainability of the national debt. When the government's debt is managed sustainably, it can respond to economic shocks, invest in public goods, and achieve its policy goals without jeopardizing its long-term financial stability. On the other hand, excessive and unmanageable debt levels significantly reduce fiscal space and a government's capacity to take appropriate action.

References

1. Australian Treasury. Economic response to the coronavirus. Australian Treasury; 2020.
2. Baldwin R, di Mauro BW, editors. Economics in the time of COVID-19. Centre for Economic Policy Research; 2020a.
3. Baldwin R, di Mauro BW, editors. Mitigating the COVID economic crisis: act fast and do whatever it takes. CEPR Press; 2020b.
4. Baldwin R, Tomiura E. Thinking ahead about the trade impact of COVID-19. In: Baldwin R, Weder di Mauro B, editors. Economics in the time of COVID-19. CEPR Press; 2020. p. 59–71.
5. Balleer A, Gehrke B, Lechthaler W, Merkl C. Does short-time work save jobs? A business cycle analysis. Eur Econ Rev. 2020;127:103442.
6. Bank for International Settlements (BIS). Annual economic report 2021. Bank for International Settlements (BIS); 2021.
7. Bank for International Settlements (BIS). Annual economic report 2022. Bank for International Settlements (BIS); 2022.
8. Blanchard OJ. Suggestions for a new set of fiscal indicators, OECD Department of Economics and Statistics working papers. OECD; 1990. p. 1–90.
9. Blanchard O, Johnson DR. Macroeconomics. 7th ed. Pearson Education; 2017.
10. Blanchard O, Pisani-Ferry J. Fiscal policy in the age of COVID-19: does debt matter? Policy Contribution, Peterson Institute for International Economics; 2020.
11. Chetty R, Friedman JN, Hendren N, Stepner M, The Opportunity Insights Team. The economic impacts of COVID-19: evidence from a new public database built using private sector data, NBER working paper no. 27431. Harvard Dataverse; 2020a.
12. Chetty R, Friedman JN, Hendren N, Stepner M, The Opportunity Insights Team. How did COVID-19 and stabilization policies affect spending and employment? A new real-time economic tracker based on private sector data, NBER working paper no. 27431. Harvard Dataverse; 2020b.
13. Chowdhury P, Paul SK, Kaisar S, Moktadir MA. COVID-19 pandemic related supply chain studies: a systematic review. Transp Res E Logist Transp Rev. 2021;148:102271.
14. Congressional Budget Office. The CARES act: estimated budgetary effects. Congressional Budget Office; 2020.
15. Congressional Budget Office. The budgetary effects of the 2020 coronavirus pandemic. Congressional Budget Office; 2021a.
16. Congressional Budget Office. The budgetary effects of major federal legislation enacted in 2020. Congressional Budget Office; 2021b.
17. Congressional Budget Office. The budgetary effects of the CARES act. Congressional Budget Office; 2021c.
18. Domar ED. The 'burden of the debt' and the national income. Am Econ Rev. 1944;34(4):798–828.
19. European Central Bank (ECB). Pandemic emergency purchase programme (PEPP). Congressional Budget Office; 2020.
20. European Commission. The EU economy after COVID-19: implications for economic governance. European Commission; 2020.
21. European Commission. NextGenerationEU: Europe's recovery plan. European Commission; 2021a. Retrieved from https://ec.europa.eu/info/strategy/recovery-plan-europe_en
22. European Commission. Policy responses to the COVID-19 pandemic. European Commission; 2021b.
23. European Commission. Recovery plan for Europe. European Commission; 2021c.
24. Federal Reserve. Policy responses to COVID-19. Federal Reserve; 2020.
25. Ferguson NM, Laydon D, Nedjati-Gilani G, et al. Impact of non-pharmaceutical interventions (NPIs) to reduce COVID-19 mortality and healthcare demand. Imperial College London; 2020.
26. Furceri D, Loungani P, Ostry JD, Pizzuto P. Will COVID-19 have long-lasting effects on inequality? Appl Econ Lett. 2021;28(7):559–65.
27. Gaspar, V., Medas, P., Perrelli, R., & Ture, H. (2020). Fiscal policy in the COVID-19 crisis: evidence from the IMF's policy tracker. International Monetary Fund working paper.
28. Gentilini U, Almenfi M, Orton I, Dale P. Social protection and jobs responses to COVID-19: a real-time review of country measures. World Bank; 2020.
29. Gentilini U, Almenfi M, Orton I, Dale P. Social protection and jobs responses to COVID-19: a real-time review of country measures. World Bank; 2021.
30. Government of Canada. COVID-19 economic response plan. Government of Canada; 2020.

31. Government of India. Pradhan Mantri Garib Kalyan Yojana. Government of India; 2020.
32. Gupta S, Scharle A, Tandon A. Fiscal policies to protect people during the COVID-19 pandemic. Center for Global Development; 2020.
33. Heller PS. Back to basics: fiscal space: what it is and how to get it. Finan Dev. 2005;42(2):46–9.
34. HM Treasury. The coronavirus job retention scheme. HM Treasury; 2020.
35. International Labour Organization (ILO). ILO monitor: COVID-19 and the world of work. International Labour Organization (ILO); 2020.
36. International Labour Organization (ILO). ILO monitor: COVID-19 and the world of work. 7th ed. International Labour Organization (ILO); 2021a.
37. International Labour Organization (ILO). World employment and social outlook: trends 2021. International Labour Organization (ILO); 2021b.
38. International Monetary Fund (IMF). Fiscal monitor: policies for the recovery. International Monetary Fund (IMF); 2020.
39. International Monetary Fund (IMF). Fiscal monitor: a fair shot. International Monetary Fund (IMF); 2021a.
40. International Monetary Fund (IMF). Digital solutions for direct cash transfers in crisis situations. International Monetary Fund (IMF); 2021b.
41. International Monetary Fund (IMF). World economic outlook, April 2021: managing divergent recoveries. International Monetary Fund (IMF); 2021c.
42. International Monetary Fund (IMF). Fiscal monitor: strengthening the credibility of public finances. International Monetary Fund (IMF); 2021d.
43. International Monetary Fund (IMF). World economic outlook: war sets back the global recovery. International Monetary Fund (IMF); 2022a.
44. International Monetary Fund (IMF). Fiscal policies to support the sustainable development goals. International Monetary Fund (IMF); 2022b.
45. International Monetary Fund (IMF). Global debt database. International Monetary Fund (IMF); 2022c.
46. International Monetary Fund (IMF). Finance & development (September 2020), a comprehensive overview of the concept.
47. IRS. IRS provides guidance on deferring employment tax payments. IRS; 2020.
48. Japanese Ministry of Internal Affairs and Communications. Special cash payment program. Japanese Ministry of Internal Affairs and Communications; 2020.
49. Keesara S, Jonas A, Schulman K. Covid-19 and health care's digital revolution. N Engl J Med. 2020;382(23):e82.
50. Keynes JM. The general theory of employment, interest and money. Macmillan; 1936.
51. Kubota K, Onishi H, Toyama T. The consumption response to COVID-19 cash transfers: evidence from a regression discontinuity design. J Econ Behav Organ. 2021;190:662–80.
52. McKinsey & Company. The COVID-19 recovery will be digital: a plan for the first 90 days. McKinsey & Company; 2020.
53. Ministry of Finance, Government of India. Economic survey 2020–21. Government of India; 2021a.
54. Ministry of Finance, Government of India. Atmanirbhar Bharat economic stimulus package. Government of India; 2021b.
55. Miroudot S. Resilience versus robustness in global value chains: some policy implications. VoxEU.org; 2020.
56. Musgrave RA, Musgrave PB. Public finance in theory and practice. McGraw-Hill; 1989.
57. Organisation for Economic Co-operation and Development (OECD). COVID-19 and fiscal relations across levels of government. OECD policy responses to coronavirus, Organisation for Economic Co-operation and Development (OECD); 2020a.
58. Organisation for Economic Co-operation and Development (OECD). COVID-19: protecting people and societies. Organisation for Economic Co-operation and Development (OECD); 2020b.
59. Organisation for Economic Co-operation and Development (OECD). COVID-19: joint actions to win the war. Organisation for Economic Co-operation and Development (OECD); 2020c.
60. Organisation for Economic Co-operation and Development (OECD). The digital transformation of government: using digital technologies to improve public sector governance. Organisation for Economic Co-operation and Development (OECD); 2020d.
61. Organisation for Economic Co-operation and Development (OECD). Tax policy reforms 2020: OECD and selected partner economies. Organisation for Economic Co-operation and Development (OECD); 2020e.
62. Organisation for Economic Co-operation and Development (OECD). The territorial impact of COVID-19: managing the crisis across levels of government. Organisation for Economic Co-operation and Development (OECD); 2020f.
63. Organisation for Economic Co-operation and Development (OECD). Economic outlook, volume 2021 Issue 2. Organisation for Economic Co-operation and Development (OECD); 2021a.
64. Organisation for Economic Co-operation and Development (OECD). The OECD economic outlook. Organisation for Economic Co-operation and Development (OECD); 2021b.
65. Organisation for Economic Co-operation and Development (OECD). Green budgeting: towards common principles. Organisation for Economic Co-operation and Development (OECD); 2021c.
66. Roy R, Heuty A. Fiscal space: policy options for financing human development. UNDP; 2009.

67. Suri T, Jack W. The long-run poverty and gender impacts of mobile money. Science. 2016;354(6317):1288–92.
68. The Conference Board. Global consumer confidence survey. The Conference Board; 2021.
69. Transparency International. Corruption and the COVID-19 pandemic: a deadly mix. Transparency International; 2021.
70. UNICEF. Social protection response to COVID-19 in South Asia. UNICEF; 2020.
71. UNIDO. Industrial development report 2022: the future of industrialization in a post-pandemic world. United Nations Industrial Development Organization; 2021.
72. United Nations (UN). The sustainable development goals report 2020. United Nations (UN); 2020a.
73. United Nations (UN). Policy brief: COVID-19 and universal health coverage. United Nations (UN); 2020b.
74. United Nations Conference on Trade and Development (UNCTAD). Trade and development report 2021. United Nations Conference on Trade and Development (UNCTAD); 2021.
75. United Nations Development Programme (UNDP). Integrating the SDGs into national budgetary processes: Rwanda's experience. United Nations Development Programme (UNDP); 2019.
76. United Nations Development Programme (UNDP). Human development report 2020: the next frontier—human development and the Anthropocene. United Nations Development Programme (UNDP); 2020a.
77. United Nations Development Programme (UNDP). SDG budget tagging: a review of early experiences. United Nations Development Programme (UNDP); 2020b.
78. United Nations Development Programme (UNDP). Digital solutions for crisis management and resilience building. United Nations Development Programme (UNDP); 2021a.
79. United Nations Development Programme (UNDP). Temporary basic income: protecting poor and vulnerable people in developing countries. United Nations Development Programme (UNDP); 2021b.
80. United Nations Development Programme (UNDP). The COVID-19 pandemic: humanity needs leadership and solidarity to defeat the coronavirus. United Nations Development Programme (UNDP); 2021c.
81. United Nations Development Programme (UNDP). The social and economic impact of COVID-19 in the Asia-Pacific region. United Nations Development Programme (UNDP); 2021d.
82. United Nations Environment Programme (UNEP). COVID-19 and human development: assessing the crisis, envisioning the recovery. United Nations Environment Programme (UNEP); 2021e.
83. United Nations World Tourism Organization (UNWTO). UNWTO world tourism barometer and statistical annex, January 2021. United Nations World Tourism Organization; 2021.
84. World Bank. Global economic prospects: recovery amid uneven progress. World Bank; 2021a.
85. World Bank. Government-to-person (G2P) payments: lessons from the COVID-19 response. World Bank; 2021b.
86. World Bank. Poverty and shared prosperity 2021: reversals of fortune. World Bank; 2021c.
87. World Bank. Social protection and jobs responses to COVID-19: a real-time review of country measures. World Bank; 2021d.
88. World Bank. World development report 2021: data for better lives. Washington, DC: World Bank; 2021e.
89. World Bank. World development report: data, digitalization, and development. World Bank; 2021f.
90. World Bank. Global economic prospects. World Bank; 2022a.
91. World Bank. The state of economic inclusion report 2022: the potential to scale. World Bank; 2022b.
92. World Health Organization (WHO). Timeline of WHO's response to COVID-19. World Health Organization (WHO); 2020a. Retrieved from https://www.who.int
93. World Health Organization (WHO). Pulse survey on continuity of essential health services during the COVID-19 pandemic. World Health Organization (WHO); 2020b.
94. World Health Organization (WHO). Global excess deaths associated with COVID-19, January 2020–December 2022. Geneva: World Health Organization; 2023.
95. World Trade Organization. World trade statistical review 2021. WTO Publications; 2021.

9 Environmental Impact of Pandemic and Potential Strategies of Sustainability

Nishant Ranjan, Nilendu Basak, and Dhruv Mishra

9.1 Introduction

The COVID-19 pandemic has impacted the whole world, it has not only led to a high number of infections and deaths which are the primary causes of concern of the virus but it also led to many secondary concerns one of which can be seen as its impact on the environment which can be taken as both positive and negative [1, 2]. Looking at the positive aspects of the pandemic on the nature, we saw a reduction in the level of pollution whether it's air, water, or noise pollution [1, 3, 4]. On the negative side, the pandemic significantly increased the generation of solid waste both organic and inorganic, along with biomedical waste, and also there was a reduced recycling of this waste to curb the spread of disease [1, 2, 5] (Fig. 9.1). The pandemic, which originated in a seafood market in Wuhan city of China, soon spread to many countries of the world due to which many countries imposed the policy of lockdown and quarantine. Among the main problems linked with this is the increase in the use of plastics as people rely mainly on online shopping and also industries take advantage of the pandemic to overturn the rules that were there to restrict the use of single-use plastics [2, 6]. Mismanagement in the appropriate dumping of these protective gears may lead to the spreading of disease and these plastics can serve as a vector of the virus, which survives upon plastics for up to 3 days [7]. Another issue associated was the upsurge in the rate of biomedical waste generation, the main contribution in this comes from the diagnostic kits, PPE kits, etc. According to reports and published data in different articles the amount of generation of this waste spiked suddenly for example according to Zambrano-Monserrate et al. the per day generation of medical waste in Wuhan, China was 240 metric tonnes on a daily basis throughout the first wave of the pandemic, nearly 190 metric tonnes more than the usual daily waste generation similarly other data like for the Indian city of Ahmedabad, generated waste during the first phase of lockdown went up to 1000 kg per day compared to previous data of 550–600 kg per day [6]. Along a similar line, the waste generation in Dhaka, Manila, Kuala Lumpur, Hanoi, and Bangkok increased nearly by 154–280 metric tonnes every day [8, 9]. There is a need for sustainable practices like there should be enhanced waste segregation to sort the waste at the source to prevent the mixing of biomedical and municipal waste

N. Ranjan (✉)
College of Basic Sciences & Humanities,
Dr. Rajendra Prasad Central Agricultural University,
Pusa, Samastipur, Bihar, India

N. Basak
Department of Microbiology, University of Kalyani,
Kalyani, West Bengal, India

D. Mishra
Department of Biological Sciences, College of Basic
Sciences and Humanities, G.B. Pant University of
Agriculture and Technology, Pantnagar,
Uttarakhand, India

A. Raghav et al. (eds.), *COVID-19 Impact on Economy, Environment and Healthcare*,
https://doi.org/10.1007/978-981-95-1196-9_9

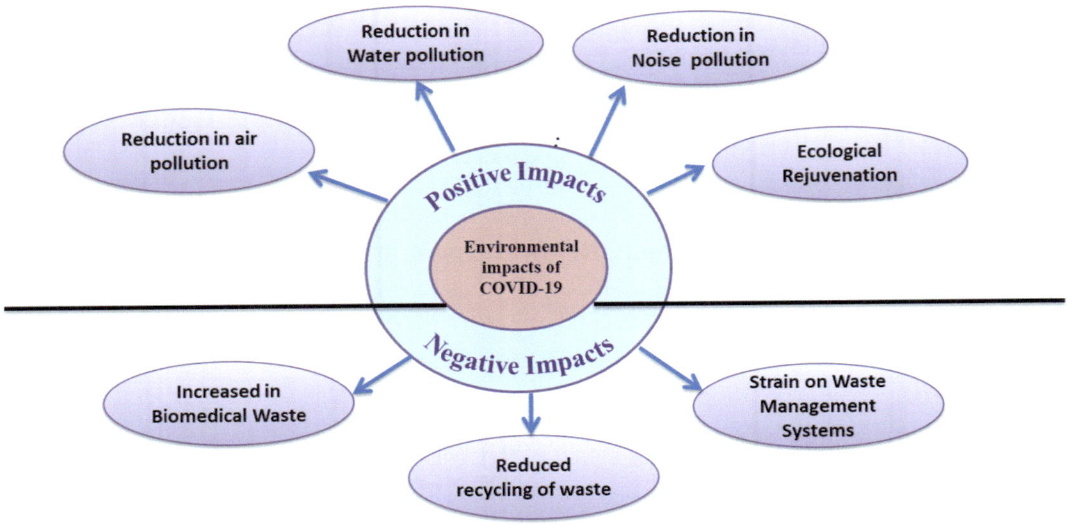

Fig. 9.1 Environmental impacts of COVID-19: positive and negative aspects

[10]. Proper incineration and careful handling of biomedical waste are needed. Moreover, there is research scope to develop reusable, washable masks, PPE, and gloves with improved antibacterial activities and specialized coatings [6]. It is important to develop awareness and educate citizens on the responsible disposal of plastic waste, especially of face masks and other single-use plastic. This approach helps to minimize the environmental damage triggered by littering and inappropriate waste disposal [11]. Also, there should be government intervention in regulating the discarding of biomedical waste and the promotion of sustainable practices [12]. In this chapter, we look at the positive and negative environmental impacts of pandemics and highlight sustainable strategies to lessen the risk of future pandemic impacts.

9.2 COVID-19: A Major Pandemic of the Twenty-First Century

On March 11th of 2020, the WHO (World Health Organisation) acknowledged the coronavirus disease 19 (COVID-19) outbreak as a pandemic. The recent pandemic, which commenced in Wuhan (China) in the last month of 2019, affected the globe and restricted movement across the world. In the twenty-first century, coronavirus is becoming a greater menace to human health being responsible for high mortality and morbidity. For the past 20 years, human coronaviruses have been identified as causative agents of mild illness [13]. Nearly two decades ago in 2003, the first outbreak was brought on by SARS-CoV (severe acute respiratory syndrome coronavirus) in the Guangdong area of China. Due to international travel, it had an impact on 20 countries. In that wave, nearly 8098 cases were registered and approximately 916 of them led to death. Even though the pandemic ended in 2004, it resulted in a global loss of nearly $90 billion [14–17]. After 10 years of SARS-CoV infection, in November 2019, the MERS-CoV (Middle East Respiratory Syndrome Coronavirus) confronted the globe. Back then, this disease impacted nearly 2494 individuals resulting in about 858 fatalities [18, 19]. The third pandemic of a novel Beta-coronavirus strain occurred at a seafood market in Wuhan, China, on 31st December 2019 [20]. This epidemic initially appeared in the Chinese population mainly linked to their common habit of eating wild animals like snakes, bats, etc. [21]. The disease was thought to be primarily zoonoses with Pangolins, Snakes, and bats as the intermediate host of the virus. Coronavirus also crossed species barrier and was transmitted from animals to infect humans. However, the link between the cross-species infection of humans from bats has no scientific evidence as of now [17, 22]. The infection is spread to humans by inhaling or coming in contact with the droplets or aerosol particles released by a diseased person. The development period of the disease ranges from 5 to 12.5 days and a maximum of 14 days. Viral aerosol particles are active for 3 h in the air, 24 h on cardboard, and 2–3 days on stainless steel and plastics. Fever, cough, dyspnoea, sore throat, and lethargy are typical symptoms of the disease. Acute respiratory distress syndrome, pneumonia, and multiple organ dysfunction can develop in some individuals. The chances of getting infected are increased if the person is immunocompromised, has heart illness, lung disease, or cancer, if the person is old, diabetic, or if the woman is pregnant [23]. The initial stage of viral infection involves the host cells' receptor binding which leads to the fusion of the cell membrane. Lung epithelial cells are believed to be the main or major target for the virus as a result of which, it was seen that the SARS-CoV is spread by connecting the virus's spike receptor binding domain to the recognized cellular receptor called the angiotensin-converting enzyme 2 receptor. Additionally, the COVID-19 virus sequence points to the receptor binding domain corresponding to one of the SARS-CoV. The data here clearly shows that ACE2 receptors are most probably the point of entrance of the virus [24–26]. According to World Health Organization data, the COVID-19 pandemic has resulted in approximately 778 million cases and 7.1 million deaths worldwide since its emergence [27].

Protective procedures suggested by the World Health Organization for prevention from the disease include:

1. Cleaning of hands regularly with alcohol-based hand sanitizer or soap.
2. Keeping a safe distance of minimum 1 m from an individual who is coughing or sneezing.
3. Avoid touching head, nose, and mouth with unclean hands.
4. When coughing or sneezing, covering mouth and nose with a bent elbow or tissue to practice respiratory hygiene.
5. If there are symptoms like fever, cough, and breathing-related issues then medical help should be taken.
6. Curbing the gathering in populous areas such as religious places, markets, and tourist places.
7. Shifting schools and colleges to online mode.

9.3 Positive Environmental Impacts of COVID-19

The pandemic triggered a global lockdown by March 2020 and changed human lifestyles notoriously, mostly removing the human impact on the environment on a huge scale. The restrictions on industrial operations, transportation, public gatherings, etc. created several positive environmental effects, some unexpected, while some obvious. But in essence, nature got a cooldown period, and the significant healing of the environment made us ponder the various impacts that we humans have on our environment. The major benefits granted to the environment during and after the lockdown were, as expected, the decrease in air, water, and noise pollutants [28]. Fewer vehicles on the roads and fewer industrial activities led to the lack of air pollution and the betterment of air quality. Similar water pollution was also decreased due to less industrial waste, thus improving the habitat of the aquatic animals. The absence of humans outside in the cities even led to occasional sightings of wild animals in cities. Even the noise was reduced to the overall lack of human activities like gatherings, traffic jams, etc. In the following sections, we look at the positive impact on the environment and learn more about reducing air, water and noise pollution. In doing so, we highlight concrete examples and data from different countries, illustrating how even a temporary change in human activities and lifestyles impacts the environment, while also discussing long-term sustainability and the policy reforms required to achieve it.

9.3.1 Reduction in Air Pollution

Due to the lesser number of automobiles on streets, a drop in the number of industries that were functioning, and also most of the companies, and institutions shifting to online mode during lockdown led to a drop in air pollution as well as greenhouse gas emissions. In India, the CPCB (Central Pollution Control Board) reported that the air quality had enhanced significantly compared to the usual circumstances owing to the decreased atmospheric concentration of PM2.5, PM10, NO, SO_2, CO, and toluene and the enhanced concentration of ozone which was primarily due to the reduced NO_2 emissions during the lockdown. There was nearly 50% less NO_2 emission during the nationwide lockdown in comparison to the normal situation, during the lockdown, there was a decline in the emission of PM2.5, PM10, NO_2, O_3, & SO_2 [29]. In Delhi, the NO_2 level dropped by 70% compared to the scenario before the COVID lockdown [30]. Similarly, in New York, compared to 2019, during the lockdown period air pollution dropped by 50% in 2020 [31]. The European Environmental Agency recorded a 60% decline in NO_2 release during the lockdown across major European cities [32]. NO_2 level also dropped from 4.5 to 1 ppb in Canada [33]. N_2O and CO emissions also dropped during the lockdown. In China, the N_2O and CO emissions dropped by 50% [34]. NO_2 is emitted by vehicle exhaust from the burning of fossil fuels. It is reported that the presence of NO_2 is related to acid rain when it interacts with O_2 and H_2O and it causes several respiratory diseases [35]. Vehicles and airplanes are considered the major source of emission of GHGs contributing 72% and 11% respectively, from the transport sector [31]. Due to restrictions imposed by countries on international travel, emissions reduced significantly. The change in air quality across the globe on different parameters is summarized in Table 9.1

9.3.2 Reduction in Water Pollution

In most developing countries, pollution of water is a major problem, as domestic and industrial waste is directly discarded into rivers without appropriate treatment [44, 45]. During the pandemic, most industries were entirely closed which was the main cause for the enhancement in the quality of water as the key reason of water

Table 9.1 Impact of COVID-19 lockdown on air quality parameters in different locations/regions

Location/region	Air quality parameters (changes observed)	References
Eastern Province, KSA	Decrease pattern in NO_2 (by 12–86% during lockdown and 14–81% post-lockdown); decrease in PM10 (by 21–70%), CO (by 5.8–55%), SO_2 (by 8.7–30%); increase in O_3 (by 6.3–45%)	[36]
Delhi, India	PM10: >50% reduction; PM2.5: >50% reduction; reduction in NO_2 (-52.68%); and CO (-30.35%)	[37]
Poland	Reductions of aerosol concentrations; Reductions of tropospheric NO_2 (approximately by -10% to 19%)	[38]
11 representative cities in Spain	Decrease in level of CO, SO_2, and PM10 in some cities; Reduction of NO_2 in all cities (except Santander); increased in O_3	[39]
Ankleshwar and Vapi, Gujarat, India	Significant reduction in level of PM10, PM2.5, CO, SO_2; decreased NO_2 (by 80.18%)	[40]
Madrid and Barcelona	NO_2 decreased by 62% (Madrid) and 50% (Barcelona)	[3]
São Paulo, Brazil	Decreased in level of NO (by up to -77.3%), NO_2 (by up to -54.3%); CO (by up to -64.8%); increased O_3 (by approximately 30%)	[41]
Wuhan, China	Decreased AQI (by 33.9%), PM2.5 (by 36.9%) and NO_2 (by 53.3%); increased ozone (O_3) (by 116.6%)	[42]
Beijing-Tianjin-Hebei (BTH), China	Reduction in AQI (by 15.2%), NO_2 (by 37.8%), PM10 (by 33.6%), PM2.5 (by 21.5%), and CO (by 20.4%)	[43]

pollution is industrial discharge. During the lockdown, major rivers like Ganga and Yamuna attained better water purity. As a matter of fact, out of 36 real-time monitoring stations, 27 stations were found within permissible limits [46]. This water quality shows that there is decreased concentration of coliform bacteria and fecal coliform bacteria in water throughout the lockdown period. Significant reduction in the level of heavy metal concentrations present in these water bodies was also seen and similarly there was a drop in the level of total dissolved salt, a reduction in the Biochemical oxygen demand, and increase in the dissolved oxygen content was seen. In Grand Canal city (Italy), the quality of water improved considerably and the appearance of many rare aquatic species was seen due to the water getting clearer. The quality of water in the beach areas of Malaysia, Thailand, India, Indonesia, and Bangladesh was enhanced due to the restrictions implemented during the lockdown. The amount of solid trash which was mainly generated from the construction sites also reduced leading to a reduction in soil and water pollution. All through the lockdown, there was a drop in the movement of merchant ships which further reduced aquatic pollution [1]. The changes in the water quality of different water bodies across the globe are summarized in Table 9.2.

9.3.3 Reduction in Noise Pollution

Increased sound levels from numerous human activities (like machinery, cars, and construction projects) are known as noise pollution, and they impact various living things in a bad way [2, 56]. As per the reports, around 360 million humans are at risk of hearing loss due to noise pollution [57]. Noise generated due to anthropogenic action may have negative impacts on wildlife as well. However, lockdown instructions directed the people to stay inside their home which led to a reduction in travelling, economic and cultural activities, and communication which eventually lowered the noise in most cities [1, 2]. In Delhi, there was a nearly 40–50% reduction in the noise level [6]. Additionally, as a result of restrictions

Table 9.2 Effect of COVID-19 lockdown on water quality of water bodies of various locations

Location/water body	Improved water quality parameters	References
Meriç-Ergene River Basin (Ergene River and Çorlu Stream), Turkey	Decrease in concentration of 'Cr', 'Ni', 'Zn', 'Cu', 'As', 'Pb', and 'Cd'; reducing pattern in heavy metal pollution index	[47]
Damodar River, India	Improvement in Modified Water Quality Index; decrease in Heavy Metal Index; shift from 'eutrophic/hyper-eutrophic' to 'oligotrophic/meso-eutrophic' condition	[48, 49]
Zarjoub River, North Iran	Reduction in concentration of 'Al', 'As', 'Ba', 'Cr', 'Cu', 'Mo', 'Ni', 'Pb', 'Se', and 'Zn'; Reduction in water pollution (by 30%); Improvement in Multi-purpose Water Quality Index (by 34%)	[4]
Chennai Coast, India	SPM decrease (15.48% in Chennai port, 37.50% in Ennore port, 22.26% in Adyar, and 33.97% in Cooum riverine estuaries); PM2.5 and PM10 reductions (24.38% and 28.43% in Adyar, 36.09% and 67.18% in Nungambakkam, and 45.63% in Ennore-Manali Industrial region)	[50]
Ganga River, India	Improvement in Dissolved Oxygen; reductions in TOC (21.28%), NO_3- (34.34%), NH_4^+ (17.84%), and PO_4^{3}- (22.06%)	[51, 52]
Bagmati River Basin, Nepal	Increase in DO levels (by 1.5 times); decrease in BOD (by 1.5) and COD (by 1.9)	[53]
River Gomti, India	Reduction in heavy metal concentrations (As, Cd, Cr, Fe, Mn, Pb)	[54]
Hooghly River, India	Reduction in dissolved heavy metals (Cd by 50%, Pb by 53%); TDS (62%); BOD (52%); total coliform (63%); faecal coliform (61%)	[48, 49]
Thirumanimuthar River, India	Reduction in microbial populations (fecal coliform, total coliform, *E. coli*, fecal streptococci) and heavy metals (Fe, Cu, Cd, Cr); improvement in Heavy Metal Pollution Index	[55]

on travel, there are significantly a smaller number of flights and vehicles internationally which has reduced noise pollution. For example, in Germany, passenger air travel was reduced by more than 90%, car traffic was slashed by 50%, and train services were 25% less than usual [57]. Overall, there was a reduction in noise pollution all over the planet because of the reduced economic activities which are due to the COVID-19 lockdown.

9.3.4 Increased Wildlife Sightings

Due to the restrictions imposed during the lockdown, wildlife has seen some enormous changes as the wild animals have explored some human-restricted areas which were previously unknown to them. An app released by the Wildlife Institute of India provided real-time movements of the animals. This app 'Lockdown Wildlife Tracker' allows people to share the movement of animals and upload them so that these data can be used later for academic and research purposes [58]. The short-term lockdown benefitted the flora and fauna of nature. Animals which typically hide in the shadow of surrounding suddenly appeared roaming in search of food. Mountain goats were spotted on the roads in Wales, UK. Similarly, a leopard was sighted on the road in Hyderabad and Peahens were spotted near M.N. Marg, New Delhi. Boars were spotted on the roads of Barcelona, which was previously dominated by vehicles. Even in Santiago (Chile), wild puma was observed. Similarly, Orcas started exploring around in North America, due to the lack of human presence. Underground moles started crawling above the ground near the footpath. These all were uncommon events which was shown during the lockdown [59]. The lockdown also affected aquatic life significantly like turtles were spotted nesting peacefully on the beaches in Florida. According to reports in Times of India Ganges dolphin returned to Kolkata after 30 years and the spotting of majestic whales was reported in parts of Vancouver for the first time in the last few decades. Also, noise pollution impacts the birds negatively, during the lockdown the birds were also seen living peacefully. Further changes were seen in the migratory behaviour of birds.

9.4 Negative Environmental Impacts of COVID-19

This pandemic has had a huge effect on waste generation and management practices on a global level, which in turn led to unprecedented challenges for public health and environmental sustainability. The result was a dramatic increment in biomedical wastage that came from various healthcare services, disinfection protocols, and widespread use of PPEs (Personal Protective Equipment). All these related events together exposed the ineffectiveness of all the waste management infrastructures and policies throughout the world. The pandemic tested the resilience of the waste management systems, whether it be the overflowing hospital waste bins, the surge in municipal solid waste that was caused by the lockdown-induced online shopping, or the resurgence of single-use plastics due to the suspension of recycling programs. As a result of this, countries came to face heightened risks of disease transmission due to untreated biomedical waste. At the same time, the strained supply chains and the panic buying have worsened the shortage of essential PPE.

9.4.1 Increase in Biomedical Waste Generation

The COVID-19 pandemic has led to a substantial rise in medical waste globally, posing serious threat to both public health and the environment [60]. This surge in waste is primarily due to the increased need of sample collection from suspected COVID-19 patients, extensive treatment protocols, and disinfection measures in hospitals, resulting in a large volume of infectious and biomedical waste [51, 52, 61, 62]. The alarming increase in COVID-19 waste is evident; for instance, Wuhan's daily output skyrocketed from

40 to 240 tonnes, while the United States experienced a rise from 5 million tonnes annually to 2.5 million tonnes monthly. By early May 2020, South Korea had generated approximately 2000 tonnes of COVID-related waste, and global demand for personal protective equipment kits (PPE kits) is projected to rise by 40% monthly, with a sustained growth rate of 20% expected until 2025 [63–65]. During the pandemic, India produced about 101 metric tonnes of biomedical waste (BMW) daily, in addition to the pre-existing average of 609 metric tonnes each day [66]. Similarly, in Ahmedabad, India, medical waste generation jumped from approximately 550–600 kg per day to around 1000 kg daily during the initial lockdown phase [6]. In Dhaka, Bangladesh's capital, daily medical waste reached about 250 metric tonnes due to COVID-19 [67, 68], with other cities like Manila, Kuala Lumpur, Hanoi, and Bangkok reporting increases of 154–280 metric tonnes daily compared with the pre-pandemic days [8]. The increasing volume of untreated biomedical waste (BMW) poses a significant risk of disease transmission among waste handlers and nearby communities, creating unprecedented challenges for local waste management authorities in addressing this hazardous material. It is crucial to manage hospital-generated waste—such as needles, syringes, masks, gloves, used tissues, and discarded medications—effectively to mitigate further infections and environmental contamination. This issue has become a global concern that requires urgent attention.

9.4.2 Widespread Use and Mismanagement of Safety Equipment During COVID-19

During COVID-19, there is an increase in the usage of Personal Protective Equipment during the period from late 2019 to 2022 [69]. Due to the airborne nature of the pathogen, the use of a mask was suggested [70]. The use of this equipment by clinicians increased when caring for affected patients. The personal protective equipment includes a face mask (N95), a face shield, a gown, and a pair of gloves [71]. The exponential rise in the number of cases of COVID-19 leads to the surge. According to WHO, an estimated need of 89 million masks, 76 million gloves for examining, 30 million gowns, and 1.6 million goggles along with 2.9 million of hand sanitizers are required per month [11, 65]. Global health organizations like WHO and other national agencies strongly suggested the use of PPE to reduce transmission. They suggested wearing facemasks and maintaining proper hygiene, and practising appropriate respiratory protocol, such as covering mouth while coughing and sneezing [5].

Soon after the spread of the disease, the shortage of PPE has been seen in major countries as healthcare professionals need this in a larger quantity. The main responsible factors can be seen as

1. The hospital's procedure for operation promotes reducing expenses rather than keeping sufficient stock of PPE which shows the flawed budgeting models of hospitals [72].
2. Existing stock of PPE had been drained by the sudden and major demand surge from healthcare services, caused by impulse buying and stockpiling [72].
3. There has been major supply disruption across the globe as most countries banned the exports of these products to fulfil the increased demand of their own [72].
4. The fear among the public of getting infected has instigated a surge in the need for safety equipment [11].
5. These shortages pose a high risk among healthcare workers and leave the frontline workers at risk [65].

9.4.3 Increased Municipal Solid Waste and Reduced Recycling

During COVID-19 there was a surge in the use of personal protective equipment kits (PPE kits) including face mask, facial shields, gloves, etc. as discussed in Sect. 9.4.2 according to the estimates of the WHO, 89 million masks, 76 million gloves for examining, 30 million of gowns, and 1.6 million of goggles along with 2.9 million of hand sanitizers are required per month [11]. Mismanagement in the proper disposal of these protective gears may lead to the spread of SARS-CoV-2 and these plastics can serve as a vector for the virus, which survives on plastics for up to 3 days [7]. The global waste generation of medical waste increased significantly during COVID-19 which was a major risk to public health and the environment, like the amount of plastic waste used for the collection of samples for diagnosis, treatment, disinfection, and biomedical waste generated by hospitals [2, 6]. For instance, according to Zambrano-Monserrate et al. [2], the per day generation of medical waste in Wuhan, China was 240 metric tonnes every single day during the first wave of pandemic, nearly 190 metric tonnes higher than the usual daily waste generation. Similarly, other data like for the Indian city of Ahmedabad, generated waste during the first phase of lockdown went up to 1000 kg per day compared to previous data of 550–600 kg/day [6]. Along a similar line, the waste generation in Dhaka, Manila, Kuala Lumpur, Hanoi, and Bangkok increased nearly by 154–280 metric tonnes per day [8, 9]. Due to the lockdown policy introduced by many countries, there was a rise in generation of municipal solid waste both organic and inorganic. Due to the pandemic, many countries established quarantine policies which led to bigger demand for online shopping for home delivery, which eventually led to a spike in the generation of domestic waste from these packaging materials [2, 6]. Waste recycling is a proven effective way for the prevention of pollution, it also helps to save energy and conserve natural resources. However, because of the pandemic, nations restricted recycling programs to prevent disease spread. For example, being worried about the spread of disease in recycling facilities, the United States reduced nearly 46% of recycling. The United Kingdom and other European countries also forbidden infected persons from entering these recycling facilities [2, 6, 73]. Additionally, despite single-use plastic can still harbour bacteria and viruses, the industry has taken advantage of the opportunity to reverse the laws that impose the ban on these plastic bags [2].

9.5 Sustainable Strategies to Mitigate Pandemic Impacts and Prevent Future Crisis

The outbreak of COVID-19 is linked to either 'zoonotic transmission' or a laboratory accident. Thus, it highlights the urgent need to prioritize sustainable mitigation strategies and consolidation of biosafety protocols to tackle this type of outbreak [74]. However, measures such as reducing deforestation and regulating wildlife trade can mitigate the risks of virus-related outbreaks, but current strategies to address COVID-19 often overlook ecosystem-based solutions [75]. A sustainable strategy needs to be implemented directly to avoid the loss of human life and socioeconomic damage. To maintain economic resilience, it is necessary to act towards a sustainable environment and agricultural approach [76].

9.5.1 Sustainable Management Through Reducing the Anthropogenic Activity

Anthropogenic activities are major factors in the emergence and spread of viral outbreaks. Rapid industrialization and related anthropogenic activities such as deforestation, mining, agricultural expansion, urbanization, etc., are degrading natural habitats and increasing the interaction between humans–wildlife. Therefore, these may increase the risk of zoonotic infections through exposure to pathogens carried by wild animals [77]. Industrialization is the main pillar of mankind's evolution but the use of sustainable industrialization is necessary in the current state. The adaptation of biofuels and bioenergy should be more forced in industries with a specific concise area, where ecological balance would not be disturbed. Moreover, compatible industries should be built together for instance, one industrial waste might be utilized in another industry as raw materials. To meet energy demands and lower greenhouse gas emissions, renewable sources of energy, such as air, solar, water power, and even biomass can be used [78]. This will reduce the demand for fossil fuels and maintain an eco-friendly environment simultaneously, manage the energy requirement. Recovery, recycling, and reuse are the three processes from which the '3R' technique is derived. Waste can be transformed into sustainable products using these methods. These techniques are mostly used to lessen the volume of solid waste dumped in landfills [79]. However, Hajam et al. [80] proposed the '11R' strategies to mitigate the waste, while two of the more promising strategies in 11R are 'rethink' and 'raise awareness'.

Cooperative efforts are required to promote sustainable development with appropriate restrictions on human health menaces for the supervision of the COVID-19 pandemic situation. Restriction on the movement in the tourist place for a certain period should be imposed. Moreover, the use of local food items rather than packaged food should be promoted to reduce carbon emissions [1]. Improvement in daily lifestyle also reduces the chances of spreading viral diseases. Preventive measures such as washing hands, sanitization, proper usage of masks and gloves, and proper dietary intake can be applied to combat diseases. More importantly, proper guidance is needed to intake the medicine especially, antibiotics.

9.5.2 Preventing Future Crises of the COVID-19 Pandemic

The pandemic did emphasize the necessity for a plan for combating future pandemics. The scientific knowledge gained from this, recognition of preparedness, response, and resilience could be reinforced.

9.5.2.1 Strengthening Surveillance and Early Warning Systems

Robust disease surveillance and rapid response are fundamental to pandemic prevention. Advanced technologies such as machine learning via artificial intelligence can analyse global health data to detect unusual patterns and enable early outbreak identification [81]. Studies emphasize the role of genomic surveillance and sequencing in monitoring the evolution of pathogens, such as SARS-CoV-2, to identify concerning variants quickly [82, 83].

9.5.2.2 Strengthening Health Systems and Vaccine Equity

The COVID-19 pandemic highlighted the requirement for strong health care infrastructure and vaccine equity. According to a report by the WHO [65], resilient healthcare systems with well-trained staff, enough diagnostic facilities, and supply chains are an absolute necessity for effective response mechanisms against the pandemic. The strategies that help in mitigating future health crises include expanding universal health coverage and building reserve capacities in emergency healthcare workers, isolation units, and intensive care facilities. Vaccination has significantly helped contain the spread and severity of COVID-19, with mRNA vaccine technology serving as a transformative tool due to its adaptability and rapid production compared to other vaccine types [84]. However, equitable vaccine distribution remains a global challenge, despite efforts by initiatives like COVAX to achieve widespread immunity. Vaccine inequity not only prolongs pandemics but also increases the risk of viral mutation and circulation [85].

9.5.2.3 Enhancing Global Health Resilience Through Cooperation and Communication

Transboundary health threats necessitate strong international cooperation. Roles and responsibilities during health emergencies can be defined by strengthening frameworks like as the International Health Regulations (IHR) and creating a global pandemic agreement, which will allow for efficient cooperation between countries [86]. Such collaboration nurtures resource pooling, knowledge sharing, and synchronized public health interventions. The pandemic also revealed weaknesses in international supply chains and pointed to the need for diversification, local manufacturing, and strategic stockpiling of critical items such as PPE, vaccines, and antivirals to ensure timely distribution during crises [87, 88]. Besides, digital tools should be utilized for real-time inventory tracking to manage logistics efficiently. Of equal importance is the fight against misinformation, which has driven vaccine hesitancy and lowered adherence to precautionary measures [89]. Governments should be investing in clear, science-based communication strategies and working with social media platforms to overcome false information. Public trust can be built only if it involves inclusivity, transparency, and continuous community engagement. Association of international entities and local health authorities is vital to ensure data sharing and interoperability of systems.

9.6 Conclusion

The COVID outbreak was one of the major pandemics in recent times infecting a large number of the population and death of infected persons. The pandemic has stalled the normal life of people as countries-imposed lockdown and quarantine policies to curb human-to-human interaction in order to keep controllable infection and death due to the infection. However, the pandemic also highlighted the interplay of how anthropogenic activities and the health of the environment. The reduction in pollution for a short period during the lockdown showed us how we can reduce the harm to the environment through sustainable practices. People started experiencing clean air and also clean water bodies mostly due to the shutdown of industries. Due to restrictions in human movement, there was a significant effect on wildlife as well. Animals were often spotted in the cities as well as on the roads near highways. A significant reduction in the level of NO_2, SO_2, PM2.5, and PM 10 was seen during the period. Significant drop in the noise level was also seen during the period. Apart from all these positive aspects, certain problems were also experienced during the period. Simultaneous surges in biomedical and plastic waste shed light on the flaws in the waste management systems that require attention. Potential solutions for these problems include segregation of the waste at the source, advanced technology for recycling, promoting the use of reusable plastics, and enforcement of stricter policies by the government to encourage sustainability. The pandemic also highlighted the need for strong healthcare systems, fair use of resources, and global monitoring for any future emergencies to prevent any future outbreaks.

References

1. Rume T, Islam SMDU. Environmental effects of COVID-19 pandemic and potential strategies of sustainability. Heliyon. 2020;6(9). Elsevier Ltd. https://doi.org/10.1016/j.heliyon.2020.e04965.
2. Zambrano-Monserrate MA, Ruano MA, Sanchez-Alcalde L. Indirect effects of COVID-19 on the environment. Sci Total Environ. 2020;728:138813. https://doi.org/10.1016/j.scitotenv.2020.138813.
3. Baldasano JM. COVID-19 lockdown effects on air quality by NO2 in the cities of Barcelona and Madrid (Spain). Sci Total Environ. 2020;741:140353. https://doi.org/10.1016/j.scitotenv.2020.140353.
4. Haghnazar H, Cunningham JA, Kumar V, Aghayani E, Mehraein M. COVID-19 and urban rivers: effects of lockdown period on surface water pollution and quality—a case study of the Zarjoub River, North of Iran. Environ Sci Pollut Res. 2022;29(18):27382–98. https://doi.org/10.1007/s11356-021-18286-5.
5. Nzediegwu C, Chang SX. Improper solid waste management increases the potential for COVID-19 spread in developing countries. Resour Conserv Recycl. 2020;161:104947. https://doi.org/10.1016/j.resconrec.2020.104947.
6. Somani M, Srivastava AN, Gummadivalli SK, Sharma A. Indirect implications of COVID-19 towards the sustainable environment: an investigation in the Indian context. Bioresour Technol Rep. 2020;11:100491. https://doi.org/10.1016/j.biteb.2020.100491.
7. Prata JC, Silva ALP, Walker TR, Duarte AC, Rocha-Santos T. COVID-19 pandemic repercussions on the use and management of plastics. Environ Sci Technol. 2020;54(13):7760–5. https://doi.org/10.1021/acs.est.0c02178.
8. Asian Development Bank (ADB). Managing infectious medical waste during the COVID-19 pandemic. Asian Development Bank (ADB); 2020. https://www.adb.org/publications/managing-medical-waste-covid19. Accessed 14 Sept 2020.
9. Rahman MM, Bodrud-Doza M, Griffiths MD, Mamun MA. Biomedical waste amid COVID-19: perspectives from Bangladesh. Lancet Glob Health.

2020;8(10):e1262). Elsevier Ltd. https://doi.org/10.1016/S2214-109X(20)30349-1.
10. Sangkham S. Face mask and medical waste disposal during the novel COVID-19 pandemic in Asia. Case Stud Chem Environ Eng. 2020;2:100052. https://doi.org/10.1016/j.cscee.2020.100052.
11. Parashar N, Hait S. Plastics in the time of COVID-19 pandemic: protector or polluter? Sci Total Environ. 2021;759:144274. https://doi.org/10.1016/j.scitotenv.2020.144274.
12. Singh N, Tang Y, Zhang Z, Zheng C. COVID-19 waste management: effective and successful measures in Wuhan, China. Resour Conserv Recycl. 2020;163). Elsevier B.V. https://doi.org/10.1016/j.resconrec.2020.105071.
13. Bonilla-Aldana DK, Dhama K, Rodriguez-Morales AJ. Revisiting the one health approach in the context of COVID-19: a look into the ecology of this emerging disease. Adv Anim Vet Sci. 2020;8(3):234–7. https://doi.org/10.17582/journal.aavs/2020/8.3.234.237.
14. Cao WC, de Vlas SJ, Richardus JH. The severe acute respiratory syndrome epidemic in mainland China dissected. Infect Dis Rep. 2011;3(1):3–6. https://doi.org/10.4081/idr.2011.e2.
15. De Wit E, Van Doremalen N, Falzarano D, Munster VJ. SARS and MERS: recent insights into emerging coronaviruses. Nat Rev Microbiol. 2016;14(8):523–34). Nature Publishing Group. https://doi.org/10.1038/nrmicro.2016.81.
16. Hui DS, Azhar E, Madani TA, Ntoumi F, Kock R, Dar O, Ippolito G, Mchugh TD, Memish ZA, Drosten C, Zumla A, Petersen E. The continuing 2019-nCoV epidemic threat of novel coronaviruses to global health—the latest 2019 novel coronavirus outbreak in Wuhan, China. Int J Infect Dis. 2020;91:264–6). Elsevier B.V. https://doi.org/10.1016/j.ijid.2020.01.009.
17. Siddique F, Abbas RZ, Mansoor MK, Alghamdi ES, Saeed M, Ayaz MM, Rahman M, Mahmood MS, Iqbal A, Manzoor M, Abbas A, Javaid A, Hussain I. An insight into COVID-19: a 21st century disaster and its relation to immunocompetence and food antioxidants. Front Vet Sci. 2021;7). Frontiers Media S.A. https://doi.org/10.3389/fvets.2020.586637.
18. Cui J, Li F, Shi ZL. Origin and evolution of pathogenic coronaviruses. Nat Rev Microbiol. 2019;17(3):181–92). Nature Publishing Group. https://doi.org/10.1038/s41579-018-0118-9.
19. Menachery VD, Schäfer A, Burnum-Johnson KE, Mitchell HD, Eisfeld AJ, Walters KB, Nicora CD, Purvine SO, Casey CP, Monroe ME, Weitz KK, Stratton KG, Webb-Robertson BJM, Gralinski LE, Metz TO, Smith RD, Waters KM, Sims AC, Kawaoka Y, Baric RS. MERS-CoV and H5N1 influenza virus antagonize antigen presentation by altering the epigenetic landscape. Proc Natl Acad Sci USA. 2018;115(5):E1012–21. https://doi.org/10.1073/pnas.1706928115.
20. Ahmad T, Haroon, Dhama K, Sharun K, Khan FM, Ahmed I, Tiwari R, Musa TH, Khan M, Bonilla-Aldana DK, Rodriguez-Morales AJ, Hui J. Biosafety and biosecurity approaches to restrain/contain and counter SARS-CoV-2/COVID-19 pandemic: a rapid-review. Turk J Biol. 2020;44(Special issue 1):132–45. Scientific and Technical Research Council of Turkey. https://doi.org/10.3906/biy-2005-63.
21. Jalava K. First respiratory transmitted food-borne outbreak? Int J Hyg Environ Health. 2020;226). Elsevier GmbH. https://doi.org/10.1016/j.ijheh.2020.113490.
22. Dhama K, Khan S, Tiwari R, Sircar S, Bhat S, Malik YS, Singh KP, Chaicumpa W, Bonilla-Aldana DK, Rodriguez-Morales AJ. Coronavirus disease 2019-COVID-19. Clin Microbiol Rev. 2020;33(4):e00028–0. https://doi.org/10.1128/CMR.00028-20.
23. Butola LK, Ambad R, Kute PK, Jha RK, Shinde AD. The pandemic of 21st century—COVID-19. J Evol Med Dent Sci. 2020;9(39):2913–8. https://doi.org/10.14260/jemds/2020/637.
24. Fehr AR, Perlman S. Coronaviruses: an overview of their replication and pathogenesis. In: Coronaviruses: methods and protocols. New York: Springer; 2015. p. 1–23. https://doi.org/10.1007/978-1-4939-2438-7_1.
25. Jaimes JA, Millet JK, Stout AE, André NM, Whittaker GR. A tale of two viruses: the distinct spike glycoproteins of feline coronaviruses. Viruses. 2020;12(1). MDPI AG. https://doi.org/10.3390/v12010083.
26. Wu P, Hao X, Lau EHY, Wong JY, Leung KSM, Wu JT, Cowling BJ, Leung GM. Real-time tentative assessment of the epidemiological characteristics of novel coronavirus infections in Wuhan, China, as at 22 January 2020. Eurosurveillance. 2020;25(3). European Centre for Disease Prevention and Control (ECDC). https://doi.org/10.2807/1560-7917.ES.2020.25.3.2000044.
27. World Health Organization. WHO COVID-19 Dashboard. Available online: https://data.who.int/dashboards/covid19/cases; https://data.who.int/dashboards/covid19/deaths?n=o (accessed on 13 August 2025).
28. Ghosh I. The emissions impact of coronavirus lockdowns, as shown by satellites. 2020. https://www.visualcapitalist.com/coronavirus-lockdowns-emissions/. Accessed 13 Apr 2020.
29. Singh V, Mishra V. Environmental impacts of coronavirus disease 2019 (COVID-19). Bioresour Technol Rep. 2021;15). Elsevier Ltd. https://doi.org/10.1016/j.biteb.2021.100744.
30. Thiessen T. How to clean air cities could outlast COVID-19 lockdowns. 2020. https://www.forbes.com/sites/tamarathiessen/2020/04/10/how-clean-air-cities-could-outlast-covid-19-lockdowns/#292a5e866bb5. Accessed 13 Apr 2020.
31. Henriques M. Will Covid-19 have a lasting impact on the environment? BBC News. 2020, March 27. https://www.bbc.com/future/article/20200326-covid-19-the-impact-of-coronavirus-on-the-environment
32. EEA. Air pollution goes down as Europe takes hard measures to combat coronavirus. European Environmental Agency (EEA), Copenhagen. 2020.

https://www.eea.europa.eu/highlights/air-pollution-goes-down-as. Accessed 14 Apr 2020.

33. Adams MD. Air pollution in Ontario, Canada during the COVID-19 state of emergency. Sci Total Environ. 2020;742:140516. https://doi.org/10.1016/j.scitotenv.2020.140516.
34. Biswal A, Singh T, Singh V, Ravindra K, Mor S. COVID-19 lockdown and its impact on tropospheric NO2 concentrations over India using satellite-based data. Heliyon. 2020;6:04764. https://doi.org/10.1016/j.heliyon.2020.e04764.
35. USEPA. Nitrogen dioxide (NO2) pollution. 2016. https://www.epa.gov/no2-pollution/basic-information-about-no2. Accessed 21 Apr 2020.
36. Anil I, Alagha O. The impact of COVID-19 lockdown on the air quality of Eastern Province, Saudi Arabia. Air Qual Atmos Health. 2021;14(1):117–28. https://doi.org/10.1007/s11869-020-00918-3.
37. Mahato S, Pal S, Ghosh KG. Effect of lockdown amid COVID-19 pandemic on air quality of the megacity Delhi, India. Sci Total Environ. 2020;730:139086. https://doi.org/10.1016/j.scitotenv.2020.139086.
38. Filonchyk M, Hurynovich V, Yan H. Impact of Covid-19 lockdown on air quality in the Poland, Eastern Europe. Environ Res. 2021;198:110454. https://doi.org/10.1016/j.envres.2020.110454.
39. Briz-Redón Á, Belenguer-Sapiña C, Serrano-Aroca Á. Changes in air pollution during COVID-19 lockdown in Spain: a multi-city study. J Environ Sci. 2021;101:16–26. https://doi.org/10.1016/j.jes.2020.07.029.
40. Nigam R, Pandya K, Luis AJ, Sengupta R, Kotha M. Positive effects of COVID-19 lockdown on air quality of industrial cities (Ankleshwar and Vapi) of Western India. Sci Rep. 2021;11(1):4285. https://doi.org/10.1038/s41598-021-83393-9.
41. Nakada LYK, Urban RC. COVID-19 pandemic: impacts on the air quality during the partial lockdown in São Paulo state, Brazil. Sci Total Environ. 2020;730:139087. https://doi.org/10.1016/j.scitotenv.2020.139087.
42. Lian X, Huang J, Huang R, Liu C, Wang L, Zhang T. Impact of city lockdown on the air quality of COVID-19-hit of Wuhan city. Sci Total Environ. 2020;742:140556. https://doi.org/10.1016/j.scitotenv.2020.140556.
43. Wang J, Xu X, Wang S, He S, He P. Heterogeneous effects of COVID-19 lockdown measures on air quality in Northern China. Appl Energy. 2021;282:116179. https://doi.org/10.1016/j.apenergy.2020.116179.
44. Islam SMD, Huda ME. Water pollution by industrial effluent and phytoplankton diversity of Shitalakhya River, Bangladesh. J Sci Res. 2016;8(2):191–8.
45. Yunus AP, Masago Y, Hijioka Y. COVID-19 and surface water quality: improved lake water quality during the lockdown. Sci Total Environ. 2020;731:139012.
46. Singhal S, Matto M. COVID-19 lockdown: a ventilator for rivers. In: Somani M, et al., editors. Bioresource technology reports, vol. 11. DownToEarth; 2020. p. 100491. https://www.downtoearth.org.in/blog/covid-19-lockdown-aventilator-for-rivers-70771.
47. Tokatli C, Varol M. Impact of the COVID-19 lockdown period on surface water quality in the Meriç-Ergene River Basin, Northwest Turkey. Environ Res. 2021;197:111051. https://doi.org/10.1016/j.envres.2021.111051.
48. Chakraborty B, Bera B, Adhikary PP, Bhattacharjee S, Roy S, Saha S, Ghosh A, Sengupta D, Shit PK. Positive effects of COVID-19 lockdown on river water quality: evidence from River Damodar, India. Sci Rep. 2021a;11(1):20140. https://doi.org/10.1038/s41598-021-99689-9.
49. Chakraborty S, Sarkar K, Chakraborty S, Ojha A, Banik A, Chatterjee A, Ghosh S, Das M. Assessment of the surface water quality improvement during pandemic lockdown in ecologically stressed Hooghly River (Ganges) Estuary, West Bengal, India. Mar Pollut Bull. 2021b;171:112711. https://doi.org/10.1016/j.marpolbul.2021.112711.
50. Vijay Prakash K, Geetha Vimala CS, Preethi Latha T, Jayaram C, Nagamani PV, Laxmi CNV. Assessment of water quality along the Southeast coast of India during COVID-19 lockdown. Front Mar Sci. 2021;8:659686. https://doi.org/10.3389/fmars.2021.659686.
51. Singh E, Kumar A, Mishra R, Kumar S. Solid waste management during COVID-19 pandemic: recovery techniques and responses. Chemosphere. 2022a;288:132451. https://doi.org/10.1016/j.chemosphere.2021.132451.
52. Singh M, Pandey U, Pandey J. Effects of COVID-19 lockdown on water quality, microbial extracellular enzyme activity, and sediment-P release in the Ganga River, India. Environ Sci Pollut Res. 2022b;29(40):60968–86. https://doi.org/10.1007/s11356-022-20243-9.
53. Pant RR, Bishwakarma K, Rehman Qaiser FU, Pathak L, Jayaswal G, Sapkota B, Pal KB, Thapa LB, Koirala M, Rijal K, Maskey R. Imprints of COVID-19 lockdown on the surface water quality of Bagmati river basin, Nepal. J Environ Manag. 2021;289:112522. https://doi.org/10.1016/j.jenvman.2021.112522.
54. Khan R, Saxena A, Shukla S. Assessment of the impact of COVID-19 lockdown on the heavy metal pollution in the River Gomti, Lucknow city, Uttar Pradesh, India. Environ Qual Manag. 2022;31(4):41–9. https://doi.org/10.1002/tqem.21746.
55. Karunanidhi D, Aravinthasamy P, Subramani T, Setia R. Effects of COVID-19 pandemic lockdown on microbial and metals contaminations in a part of Thirumanimuthar River, South India: a comparative health hazard perspective. J Hazard Mater. 2021;416:125909. https://doi.org/10.1016/j.jhazmat.2021.125909.
56. Goines L, Hagler L. Noise pollution: a modern plague. South Med J. 2007;100(3):287–94.
57. Sims J. Will the world be quieter after the pandemic? 2020. https://www.bbc.com/future/article/20200616-will-the-world-be-quieter-after-the-pandemic. Accessed 14 July 2020.

58. Wild Life Institute of India. Lockdown wildlife tracker. 2020. https://wii.gov.in/. Retrieved on 23 Apr 2020.
59. Bar H. COVID-19 lockdown: animal life, ecosystem and atmospheric environment. Environ Dev Sustain. 2021;23:8161–78. https://doi.org/10.1007/s10668-020-01002-7.
60. Sarkodie SA, Owusu PA. Impact of COVID-19 pandemic on waste management. Environ Dev Sustain. 2021;23(5):7951–60. https://doi.org/10.1007/s10668-020-00956-y.
61. Ilyas S, Srivastava RR, Kim H. Disinfection technology and strategies for COVID-19 hospital and bio-medical waste management. Sci Total Environ. 2020;749:141652. https://doi.org/10.1016/j.scitotenv.2020.141652.
62. Teymourian T, Teymoorian T, Kowsari E, Ramakrishna S. Challenges, strategies, and recommendations for the huge surge in plastic and medical waste during the global COVID-19 pandemic with circular economy approach. Mater Circ Econ. 2021;3:1–14. https://doi.org/10.1007/s42824-021-00020-8.
63. ESCAP-United Nation Economic and Social Commission for Asia and the Pacific. The safe waste treatment for COVID-19, lessons from the Republic of Korea. 2020. https://www.unescap.org/sites/default/files/200514%20waste%20management%20for%20COVID19%28edited%29%20FINAL.pdf
64. Market Reports. Personal protective equipment market by type (hands & arm protection, protective clothing, foot & leg protection, respiratory protection, Head S. Ilyas et al. Science of the Total Environment 749 (2020) 141652 11 protection), end-use industry (manufacturing, construction, oil & gas, healthcare) global forecast to 2022. 2019. https://www.marketsandmarkets.com/Market-Reports/personal-protective-equipment-market-132681971.html
65. WHO. Shortage of personal protective equipment endangering health workers worldwide. 2020. https://www.who.int/news-room/detail/03-03-2020-shortage-of-personal-protective-equipment--endangeringhealth-workers-worldwide
66. Consolidated Status Report in the matter of O.A. NO. 72 of 2020. In re: scientific disposal of bio-medical waste arising out of Covid-19 treatment—compliance of BMWM Rules, 2016 before Hon'ble National Green Tribunal, Principle Bench, New Delhi. Central Pollution Control Board, India, 2020
67. Rayhan MRI, Liza JM, Rahman MM. Assessment of COVID-19 vaccination-related medical waste management practices in Bangladesh. PLoS One. 2022;17(8):e0273053. https://doi.org/10.1371/journal.pone.0273053.
68. Sharma R, Chakraborty P, Barman SR. Decontamination strategies and technologies for tackling COVID-19 hospitals and related biomedical waste. In: Modern approaches in waste bioremediation: environmental microbiology. Cham: Springer; 2023. p. 57–99. https://doi.org/10.1007/978-3-031-24086-7_4.
69. Cubas ALV, Moecke EHS, Provin AP, Dutra ARA, Machado MM, Gouveia IC. The impacts of plastic waste from personal protective equipment used during the COVID-19 pandemic. Polymers. 2023;15(15). Multidisciplinary Digital Publishing Institute (MDPI). https://doi.org/10.3390/polym15153151.
70. Battista RA, Ferraro M, Piccioni LO, Malzanni GE, Bussi M. Personal protective equipment (ppe) in covid 19 pandemic: related symptoms and adverse reactions in healthcare workers and general population. J Occup Environ Med. 2021;63(2):E80–5. https://doi.org/10.1097/JOM.0000000000002100.
71. Ruskin KJ, Ruskin AC, Musselman BT, Harvey JR, Nesthus TE, O'Connor M. COVID-19, personal protective equipment, and human performance. Anesthesiology. 2021;134(4):518–25. https://doi.org/10.1097/ALN.0000000000003684.
72. Cohen J, Rodgers, Y van der M. Contributing factors to personal protective equipment shortages during the COVID-19 pandemic. Prev Med. 2020;141). Academic Press Inc. https://doi.org/10.1016/j.ypmed.2020.106263.
73. Ma B, Li X, Jiang Z, Jiang J. Recycle more, waste more? When recycling efforts increase resource consumption. J Clean Prod. 2019;206:870–7. https://doi.org/10.1016/j.jclepro.2018.09.063.
74. Coccia M. Meta-analysis to explain unknown causes of the origins of SARS-COV-2. Environ Res. 2022;211:113062. https://doi.org/10.1016/j.envres.2022.113062.
75. Dobson AP, Pimm SL, Hannah L, Kaufman L, Ahumada JA, Ando AW, Bernstein A, Busch J, Daszak P, Engelmann J, Kinnaird MF, Li BV, Loch-Temzelides T, Lovejoy T, Nowak K, Roehrdanz PR, Vale MM. Ecology and economics for pandemic prevention. Science. 2020;369(6502):379–81. https://doi.org/10.1126/science.abc3189.
76. Cawthorn D-M, Kennaugh A, Ferreira SM. The future of sustainability in the context of COVID-19. Ambio. 2021;50(4):812–21. https://doi.org/10.1007/s13280-020-01430-9.
77. Nazir R, Ali J, Rasul I, Widemann E, Shafiq S. Eco-environmental aspects of COVID-19 pandemic and potential control strategies. Int J Environ Res Public Health. 2021;18(7):3488. https://doi.org/10.3390/ijerph18073488.
78. Ellabban O, Abu-Rub H, Blaabjerg F. Renewable energy resources: current status, future prospects and their enabling technology. Renew Sust Energ Rev. 2014;39:748–64. https://doi.org/10.1016/j.rser.2014.07.113.
79. Snehalata A. 3R processing for different types of solid waste management. In: Biological processing of solid waste. CRC Press; 2019.
80. Hajam YA, Kumar R, Kumar A. Environmental waste management strategies and vermi transformation for sustainable development. Environ Chall. 2023;13:100747. https://doi.org/10.1016/j.envc.2023.100747.

81. Hu W-H, Sun H-M, Wei Y-Y, Hao Y-T. Global infectious disease early warning models: an updated review and lessons from the COVID-19 pandemic. Infect Dis Model. 2024; https://doi.org/10.1016/j.idm.2024.12.001.
82. Andersen KG, Rambaut A, Lipkin WI, Holmes EC, Garry RF. The proximal origin of SARS-CoV-2. Nat Med. 2020;26(4):450–2. https://doi.org/10.1038/s41591-020-0820-9.
83. Saravanan KA, Panigrahi M, Kumar H, Rajawat D, Nayak SS, Bhushan B, Dutt T. Role of genomics in combating COVID-19 pandemic. Gene. 2022;823:146387. https://doi.org/10.1016/j.gene.2022.146387.
84. Al Fayez N, Nassar MS, Alshehri AA, Alnefaie MK, Almughem FA, Alshehri BY, Alawad AO, Tawfik EA. Recent advancement in mRNA vaccine development and applications. Pharmaceutics. 2023;15(7):Article 7. https://doi.org/10.3390/pharmaceutics15071972.
85. Asundi A, O'Leary C, Bhadelia N. Global COVID-19 vaccine inequity: the scope, the impact, and the challenges. Cell Host Microbe. 2021;29(7):1036–9. https://doi.org/10.1016/j.chom.2021.06.007.
86. Jee Y. WHO International Health Regulations Emergency Committee for the COVID-19 outbreak. Epidemiol Health. 2020;42:e2020013. https://doi.org/10.4178/epih.e2020013.
87. Ranney ML, Griffeth V, Jha AK. Critical supply shortages—the need for ventilators and personal protective equipment during the Covid-19 pandemic. New Engl J Med. 2020;382(18):e41. https://doi.org/10.1056/NEJMp2006141.
88. Sachs JD, Karim SSA, Aknin L, Allen J, Brosbøl K, Colombo F, Barron GC, Espinosa MF, Gaspar V, Gaviria A, Haines A, Hotez PJ, Koundouri P, Bascuñán FL, Lee J-K, Pate MA, Ramos G, Reddy KS, Serageldin I, et al. The Lancet Commission on lessons for the future from the COVID-19 pandemic. Lancet. 2022;400(10359):1224–80. https://doi.org/10.1016/S0140-6736(22)01585-9.
89. Zang S, Zhang X, Xing Y, Chen J, Lin L, Hou Z. Applications of social media and digital technologies in COVID-19 vaccination: scoping review. J Med Internet Res. 2023;25(1):e40057. https://doi.org/10.2196/40057.

Preparing for Future Pandemic: Challenges for Healthcare, Economic, and Environmental Leadership

10

Deepanshu, Jashanpreet Kaur, Rakhi Pooja, Janvi Tandon, Aarti Aggarwal, and Anjali Saini

10.1 Introduction: The Global Challenge of Pandemic

The world has time and again learned lessons from massive outbreaks which bring about social revolutions that disrupt socio-economic and biome socio-logarithmic structures and cycles. Epidemics such as Spanish Flu, HIV/AIDS, and COVID-19 highlighted weaknesses in health system and governance structures. Moreover, pandemics such as the Spanish Flu which resulted in high mortality rate brought out the need for implementation of early measures, HIV/AIDs which is still a large and ongoing epidemic underlined the need for recurrent health investments and the need to remove the stigma attached to the disease [1]. COVID-19 unbundled interconnectivity of global systems, which highlighted variations in healthcare, economic, and environmental vulnerabilities and opportunities. From this point of view, COVID-19 brought unforeseen challenges, especially for the Indian nation where healthcare delivery systems were overstressed [2]. With reference to the current scenario we come to the problem of congestion in urban hospitals and lack of adequate facility in rural setting. The economic effects were also similar, an increase in the levels of unemployment, rise of poverty, disruptions in many sectors especially the informal business. The COVID-19 pandemic around the world demonstrated that international supply chains are weak and the imperative of developing local production capacities. Other negative impacts of the pandemic that relate to the environment are for instance enhanced medical waste production, deforestation that is associated with the emergence of zoonotic diseases [3].

This study presents a complex picture of pandemics and their implications and gives detailed recommendations for overcoming them. Solutions in healthcare are increasing the capacity, adopting digital health platforms, and a fair distribution of vaccines [4]. We also need to have the ability to move to the next sector as we see

Authors "Deepanshu", "Jashanpreet Kaur", "Rakhi Pooja" and "Janvi Tandon" have equally contributed to this chapter.

Deepanshu (✉) · A. Aggarwal
Department of Nuclear Medicine, Post Graduate Institute of Medical Education and Research, Chandigarh, India

J. Kaur · R. Pooja
Department of Endocrinology, Post Graduate Institute of Medical Education and Research, Chandigarh, India

J. Tandon
Amity Institute of Molecular Medicine & Stem Cell Research, Amity University Noida, Noida, Uttar Pradesh, India
e-mail: janvi.tandon1@s.amity.edu

A. Saini
Department of Paediatrics, Post Graduate Institute of Medical Education and Research, Chandigarh, India

A. Raghav et al. (eds.), *COVID-19 Impact on Economy, Environment and Healthcare*,
https://doi.org/10.1007/978-981-95-1196-9_10

with diversifying industries, moving to digital services, and implementing sound social protection systems. In environmental field combating climate change, increasing the protection of ecosystems and improving the process of management of wastes in the future are crucially important [5].

The experience of the Indian state during the outbreak of COVID-19 is presented in this paper as an example for some general tendencies resulting from the situation which requires not only international collaboration but also the search for additional specific approaches at the national level. Pandemic readiness requires enhancement of health systems, economic strengthening, and incorporation of the green economy principle. In other words, when societies are knowledgeable of the past pandemic and flexible to organizational deficiency, they create a strong and just world, capable of reducing the effects of the disease [4].

10.2 Lessons from Historical Pandemic

Pandemics of the past teach the possibility of the occurrence. Pandemics in the past offer great lessons which can be used when developing the right response to the health crises in the future. Every pandemic has provided very important lesson about the frailty of human and ability to survive.

1. Spanish Flu (1918–1919): This pandemic most clearly illustrated the essence of timely interventions like use of quarantine, prevention of crowd assembling, and publicity [6]. However, at that time due to the poor knowledge of medicine and the lack of vaccines, the situation was exacerbated. As shown in Table 10.1, Spanish Flu revealed measures that are supposed to receive widespread public health and the use of multilateralism [7].
2. The HIV/AIDS epidemic highlights the ongoing need for accessible health services, widespread education, and safe spaces for community support. Although advances have been made in the treatment of opportunistic infections such as antiretroviral therapy (ART), patient still experiences stigma. They discovered that this epidemic required education and strong long-term investments in health facilities, as mentioned in Table 10.1 [8].
3. COVID-19 Pandemic (2019–2023): COVID-19 exposed the essence of globalization because no country is immune to the effects of a virus from another country. It highlighted the gaps in healthcare and brought out logistic chaining, as well as shown the significance of equitable distribution of vaccines. As mentioned in Table 10.1 cohesion in the development of vaccines for COVID-19 proved the possibility of international cooperation, and the inequality in the distribution

Table 10.1 Comparative impacts of major pandemic [8]

Pandemic	Period	Death toll (millions)	Key lessons
Spanish flu	1918–1919	50	Need for timely public health measures.
HIV/AIDS	1980s–present	36	Long-term care, education, and stigma reduction.
COVID-19	2019–2023	6+	Vaccine equity, global supply chain resilience.

Table 10.2 Key impacts of pandemic across sectors [12]

Sector	Challenges highlighted	Lessons learned
Healthcare	Resource shortages, workforce burnout	Scale up infrastructure, invest in digital tools
Economy	Job losses, disrupted supply chains	Strengthen informal sector safety nets
Environment	Increased medical waste, zoonotic disease risks	Promote sustainable practices

of vaccines also revealed the existing difficulties in globalization [9].

10.3 Pandemic as a Test of Resilience

Mishaps also identify the loopholes within the society to which the health sector and the financial sectors are not exemptions. COVID-19, for example, caused the collapse of the healthcare industry interfering with the functioning of hospitals and tiring up medical staff. In Indian metro-urban-based hospitals there was acute shortage of oxygen and ventilators, at the same time rural population of India was laid helpless for primary medical treatment of minor ailments. These challenges highlighted the need for solution-centred healthcare and efficient management of resources as shown in Table 10.2. Economically, pandemics affect operations of most industries, decrease expenditure by individuals or companies, and hike unemployment rates. Consequently, the sector, which constitutes a large number of the Indian population, has been the hardest hit by the lockdown and restrictions around the globe. This is associated with mass people losing their jobs, increasing poverty, and deepening social gaps. Women and several sectors of the population were even worse off because of the low employment rates, and limited access to financing [10]. Worldwide observed supply chain disruptions led to the scarcity of goods, as mentioned in Table 10.2 which raised the awareness of the interconnectedness of countries. Small businesses especially in developing world were among the worst affected therefore there is need to adopt policies favouring economic planning and saving measures for small business among others. Environmentally, pandemics also serve to increase pollution, destruction of habitats & species, among others. The COVID-19 that originated from a wet market, a zoonotic disease, underscored the dangers of deforestation and wildlife trade, which bring the human species closer to viruses. The increase in the pandemic's medical waste includes faced masks, gloves, and test kits that overwhelmed waste management systems. Most of the developing countries struggled to manage such materials and their disposal created more problems to the environment and health of people. Moreover, the pandemic helped reduce some environmental indicators for some time—factories and industrial sites emitted less greenhouse gases, and air pollution in large cities decreased, but it was only for a short time [11].

Socially, pandemics are usually amplified by deepening social inequality and suffering impact on vulnerable groups. In COVID-19 times, the absence of digital facilities made education more unequal for students in poor families since the schools went online. The COVID-19 pandemic has resulted in more caregiving work being done by women, especially in developing countries; women's domestic safety was also threatened during lockdowns. Such social effects confirm

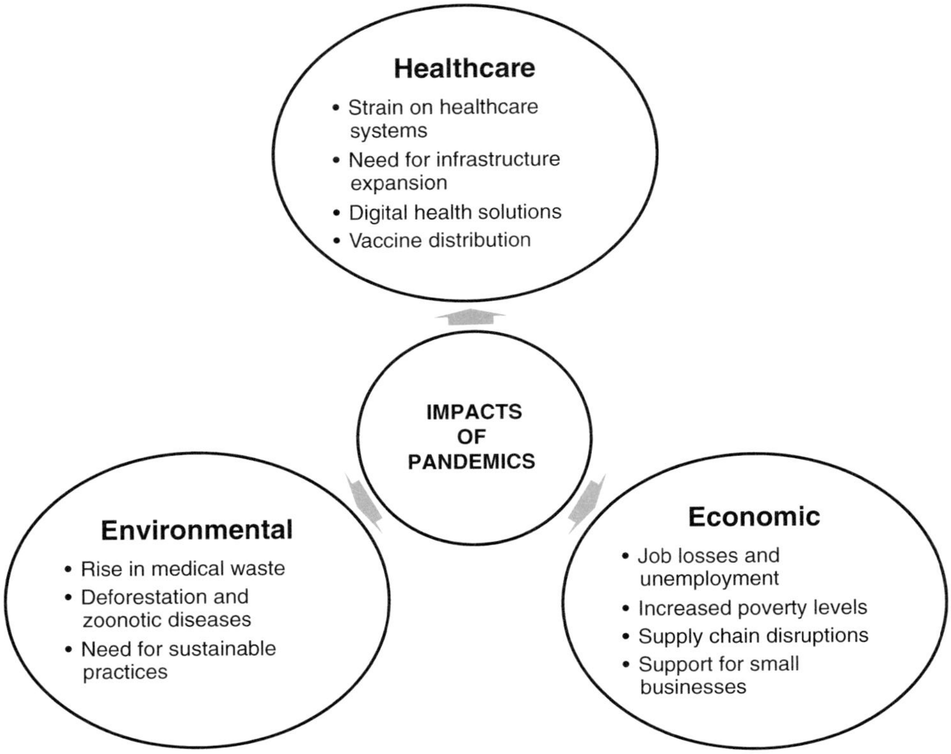

Fig. 10.1 Impact of pandemics on healthcare, environment and economic [13]

that gender, education, and community resilience as cross-cutting areas should receive attention from the authorities (Fig. 10.1).

10.4 Case Study: India's Migrant Workforce During COVID-19

The shutdowns brought upon by COVID-19 in India reduced millions of migrant workers to employed status with no work, food, or homes. Vulnerable people were the worst affected; many trekked long distances return to their villages on foot in distances without necessary arrangements. The crisis highlighted the inapt urban planning, lack of social welfare and more so inadequate means in disaster response, as shown in Table 10.3 [1].

However, pandemics also present optimisms once again as a way to reflect about as well as reconstruct systems. The fast advancement and discovery of COVID-19 vaccines demonstrated great possibilities in scientific advancement and cooperation as illustrated in Table 10.3. The community appreciation activities showed that the community can play an important role in people's

Table 10.3 Opportunities for systemic improvement post-pandemic [15]

Area	Opportunity	Example
Healthcare	Expand telemedicine and remote care	Digital health platforms in rural areas
Economy	Accelerate digital transformation	Growth of remote work and e-commerce
Environment	Strengthen focus on sustainability	Investments in renewable energy and green jobs
Social resilience	Address inequalities through inclusive policies	Gender-sensitive recovery programs

adherence to such practices as vaccination and the use of face masks. Economically, COVID-19 has advanced the technological shift and enabled work from home systems and other online purchasing platforms outlined in Table 10.3. Socially, the short interference in day-to-day human activities showed that world needs to embrace environmentally friendly practices and switch to green energy [14].

To sum up, pandemics are one of the best ways to evaluate the effectiveness of societies and transparent vulnerabilities and positive changes. They also make us realize that affairs of health, economy and environment are not mutually related but are interrelated. The above attributes imply that for societies to rebuild and design resilience for future shocks, it depends on filling of systematic voids and encouraging for invention [13].

10.5 India's Unique Challenges

Differentiated population density and social inequality affected COVID-19 in India [16]. Delhi and Mumbai were the hotspots of the virus because of massive population densities and the movement of the population. These cities had critical shortages of all medical consumables, hospital beds, and oxygen as stated in Table 10.4. Rural areas, which did not have health infrastructure to meet any challenge, found themselves overwhelmed with increasing number of cases [17]. This exposed the contradiction in the allocation in health facilities between the urban and rural regions as millions were left without access to proper healthcare whenever they fell ill. The economic effect was severe [18]. The spread of COVID was contained through lockdown measures that ensured everyone stayed at home, a measure that saw hundreds of companies laying off their employees and leaving many unemployed most of whom worked for the informal the migrant workers who could not avail any support structure were struggling a lot. This forced many to go back to their villages, they failed to be transported through cars or receive any form of financial assistance. This exodus exposed the lacuna that exist in urban planning, labour laws and social protection mechanism illustrated in Table 10.4. The women were the worst hit as most of them were laid off their jobs and or found themselves burdened with child care responsibilities. Numerous women working in cleanings services or low wage employment sectors had no income sources and poor families became poorer. Gender sensitive policies in COVID-19 response and recovery and protection of those who are in the vulnerable groups remain a burning issue [19].

Table 10.4 India's key challenges during COVID-19 [21]

Sector	Specific challenges
Healthcare	Shortages of beds, oxygen, and medical supplies.
Economy	Job losses, poverty, and labour migration.
Environment	Medical waste management, zoonotic risks.

Environmental strains, on the other hand, were also observed during the pandemic deepening the plight of peoples in the countries. Unsafe means of disposing masks, gloves, and syringes which are used in medical activities exposed threats to the well-being of people and nature highlighted in Table 10.4. Facilities responsible for waste management were unable to cope with the new volume of hazardous content especially in big cities. Some studies suggested that due to shutdowns, both water and air quality improved momentarily because industrial production was halted during the lockdowns. Pandemic showed that Governments need to adapt more friendly environmental policies in order to decrease population's vulnerability for getting zoonotic diseases and to minimize pollution [20].

10.6 Educational Disruption

The pandemic also disrupted education across India, especially for students from low-income families. With schools closed and learning shifted online, many children lacked access to the necessary digital tools, such as smartphones or computers, and reliable internet connectivity as outlined in Table 10.5. This digital divide widened

Table 10.5 Impact of COVID-19 on education in India [22]

Group	Specific challenges
Low-income families	Lack of devices, poor internet connectivity.
Girls	Increased dropout rates due to caregiving duties.
Rural students	Limited access to digital infrastructure.

Table 10.6 Social challenges and responses during COVID-19 [12]

Challenge	Response
Overcrowding in slums	Community-based distribution of sanitation kits.
Mental health issues	Increased availability of virtual counselling.
Lack of food security	Grassroots food distribution initiatives.

existing educational inequalities, particularly in rural and marginalized communities. Girls faced an additional burden, as they were often pulled out of school to assist with household chores, further reducing their opportunities for education [4].

10.7 Social Impacts and Community Resilience

Globally socially, COVID-19 deepened the existing inequalities and special disadvantage. That in urban slums infection rate was higher than in other areas because living conditions were congested and there were low numbers of available and functional sanitation facilities [23]. The pandemic exacted high stress as people lost their jobs, had loneliness, and did not know what was in store for them in terms of the illness; this led to anxiety and probably depression in many as shown in Table 10.6. Many frontline workers including the healthcare providers and other essential service providers also experienced burnout and mental fatigue. On the other hand, the pandemic also exposed

Table 10.7 Technological impacts during COVID-19 [12]

Sector	Adaptation	Challenges
Healthcare	Telemedicine for remote consultations	Limited digital literacy in rural areas.
Workplace	Shift to remote work	Lack of access to reliable internet.
Retail and payments	Growth of e-commerce and digital payments	Exclusion of non-tech-savvy populations.

the tenacity of society mentioned in Table 10.6. Local non-governmental organizations and local state departments ensured that they donate food stuff, medicals, and other basic needs to those affected. New opportunities for mental health service delivery and support during the pandemic appeared mainly due to technology use for telemedicine approach and virtual counselling [7].

10.8 Infrastructure and Technological Adaptation

COVID-19 increased the use of technology in different areas. Telemedicine for delivery of healthcare became important when physical hospitals were inaccessible due to restrictions as stated in Table 10.7. The digitalization also affected working places so that a big number of offices implemented the policy of 'work from home'. E-commerce and digital payment systems also increased sharply to indicate changing trends of consumption patterns as detailed in Table 10.7. However, such rapid technological transformation also revealed the absence of the digital divide and digital readiness. These limitations meant that a lot of small businesses and individuals could not make the leap to digital modes of operation due to lack of skills or resources hence the need for specific training's donning and development of digital facilities [18].

10.9 Future Directions

The COVID-19 pandemic engendered many instructional experienced lessons on how to tackle the system issues and how to ensure the sector's future resilience. Hence, it relates to the need for health, a better quality of population structures, and efficient urbanization with inclusive growth, among others. Sustainability in environmental practices and closing the digital divide to get the deprived section of the population back to school and earn their livelihood are now mandatory for recovery, and policies that must effectively target the weaker sections of the society including women, children, and migrant workers [18]. Improving the framework of waste disposal, recycling, and other eco-friendly solutions in the power sector will remain important in managing future shocks. By adopting these strategies in parallel fashion, India can strengthen its society through a process of preparing it for the future pandemic that might happen [24] (Fig. 10.2).

10.10 The Need for Preparedness

The cyclical nature of pandemics cannot be overstated, as they require attention to health, economic, and environmental correlates, among other areas, in order to safely protect the well-being of all inhabitants, whether locally, nationally, or internationally [16]. The finest preventative approaches are needed to address COVID-19 and other lethal diseases that have demonstrated the systems' shortcomings. In the future, frequent detection, communication, and cooperative worldwide management of such risks can significantly lessen the workload for specialists and save many lives [25].

10.11 Healthcare Preparedness

Healthcare systems must be prepared to handle a greater influx of people in a shorter amount of time in order to handle spikes in patient numbers

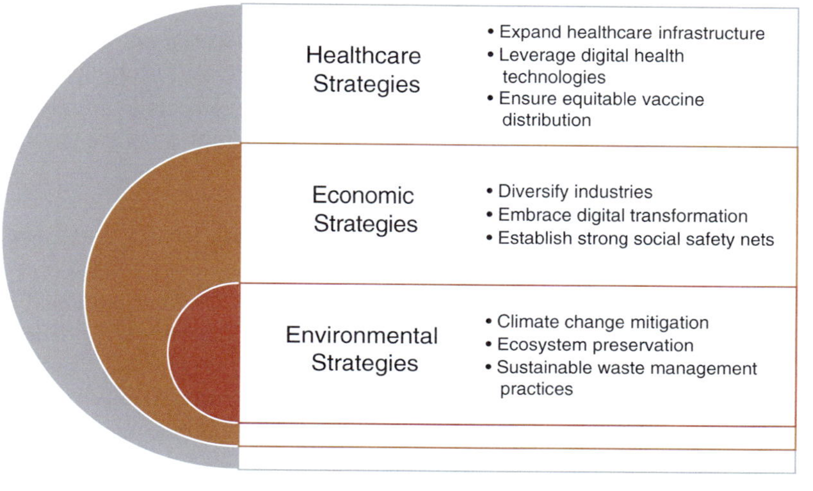

Fig. 10.2 Strategies for future preparedness [1]

during emergencies. In order to do this, hospitals must have a sufficient number of beds as well as vital medical supplies like ventilators and oxygen. In order to treat patients in a timely and suitable manner, it is also essential to increase the number of qualified healthcare personnel [11]. Access to healthcare can be greatly increased by implementing tele-health and digital health solutions, especially in rural and isolated locations. Telemedicine makes it possible for patients to see doctors without overcrowding medical facilities, which speeds up emergency diagnosis and care. To guarantee that vaccines go to the proper places during future outbreaks, governments should also concentrate on increasing their ability to produce vaccines for distribution. In order to prepare for pandemics, community-level public health must be strengthened. Programs aimed at health education, immunization campaigns, and similar initiatives can play a significant role in preventing the spread of diseases. By fostering a well-organized community response, we can better equip ourselves for future health challenges [26].

10.12 Economic Resilience

It is possible to diversify away from reliance on industries that are most susceptible to a lack of economic resilience. Another industry which has to face changed circumstances is business, therefore governments should promote innovation and digitalization [9]. For instance, it will foster social ordering platforms such as e-commerce and working from home platforms even during lockdowns. Important in implementing adequate socioeconomic policies is the establishment of provable social security measures. This encompasses issues of unemployment, unemployment compensation, financial handouts and food subsidies. It is hoped that small business owners should be offered subsidies including lower interest charges on loans and grants when the pandemic hit in order to support their recovery and operational continuation. A key focal area that needs to be advanced is the consolidation of both international and domestic supply networks. COVID-19 exposed important vulnerability of global supply chains concerning scarcity of some products such as medicines and medical equipment. Promoting local production capabilities reduces dependency on imports and ensures the availability of critical resources during emergencies [27].

10.13 Environmental Sustainability

Crises, including pandemics, make environmental consequences of actions of humans, including habitat destruction, contamination, and wildlife sales, apparent. Meeting these risks is critical to avoid the future pandemic. Thus, eliminating or limiting the risk of zoonotic diseases entering human-populated areas and preventing population growth are possible: Ecosystems should be protected, and the activity related to wildlife trade should be regulated. Proper waste management also plays an important role. Incidental disposable masks and gloves during COVID-19 strained waste management because of the increases in such waste. States need to fund that can treat most dangerous waste and minimize emissions. Mitigating factors like decarbonization processes and low-carbon food production systems are essential in preventing future pandemic. Biological health and climate stability ensure low disease risks and improve general health among a population [16].

10.14 Expanding on Strategies

1. *Public Awareness and Education:* One essential idea in preparedness is citizen education. Self-made governments should consistently communicate public health messages to the public about vaccinations, crises, and hygiene standards as exemplified in Table 10.8. People who are knowledgeable will make a special effort to follow the guidelines and support public health recommendations [26].
2. *International Collaboration:* COVID-19 falls within the category of pandemics, which are global disasters that require global reactions. Countries must produce vaccines at the same time, exchange necessities, and establish common policies for addressing pandemic

Table 10.8 Expanded strategies for comprehensive preparedness [30]

Strategy	Examples
Public awareness	Regular health campaigns on hygiene and vaccinations.
International collaboration	Sharing vaccines, establishing global health protocols.
Technological innovations	AI-based disease tracking, mobile health apps.
Rural infrastructure	Building hospitals, ensuring access to medicines.
Sustainability	Tree planting, renewable energy projects.

risks. The World Health Organization (WHO) and other international organizations serve as NSs and aid the nations [19].

3. *Technological Innovations:* In the context of pandemic response operations, technology can contribute to societal transformation. Investing in the development of AI techniques for illness monitoring and analysis as well as vaccine development might gradually increase preparedness as shown in Table 10.8. During COVID-19, health apps that alert users of potential contacts and display viral transmission rates also shown effectiveness [28].
4. *Strengthening Rural Infrastructure:* In any pandemic situation, the rural sector is in a very vulnerable position as most of these areas do not get good healthcare facilities. This means that governments should invest in what has to be described as the construction of basic facilities, in this case, the construction of healthcare facilities in the rural regions so that these regions are not neglected through the period of crisis [21].
5. *Sustainability and Green Recovery:* It would also mean that some rescue operations should have environmental objectives. Measures that include tree planting, wetlands use, and limiting carbon emissions can help to lower the impacts of climate change while at the same time boost economic development [29].

As a result, the further steps in the outlined approach will allow not only to adapt for subsequent pandemic, but also to solve the issues of public health, economic sustainability, and environment. Commitment in readiness is cyclical in nature and is a function of an organized effort, creativity regarding the endeavour of creating a world that is resilient to disasters [31].

10.15 A Holistic Approach to Pandemic Preparedness

This means that any ongoing and future work towards establishing measures against the next pandemic should encompass improvement of healthcare systems, making economic structures more adaptive, and creating economically and environmentally sound structures. The history of past pandemic especially COVID-19 does indicate the need for more preventive measures that must overcome structural vulnerabilities. To build a cooperation between the governments, International organizations, Communities, and societies have to pave the way for less vulnerable, safer, and just future [31].

10.16 Strengthening Healthcare Systems

Because healthcare systems are on the front lines in managing pandemic, it is important to explore how nations' Class F UT/OL investments are distributed. For this reason, any matter needed for the mitigation of future disasters must cover healthcare investments. This encompasses construction of new facilities, increasing the capability of existing ones and availing of the facilities with enhanced professionalism. A report on health for the people of rural areas of a country highlighted that healthcare for rural communities should be given special attention as they do not have access to quality healthcare as those in the urban areas. Telemedicine, electronic health records, and other similar tools may change the whole concept of delivering care. Telemedicine allows for virtual consultations, the overload in physical healthcare centres is avoided, and everyone, including inhabitants of distant villages, can consult a doctor. Furthermore, there is need to adopt electronic health records, because they help to plan patient care and give the healthcare providers up-to-date and accurate information [15].

At the same time, preparedness calls for a trained and motivated healthcare worker. There is need for governments to provide funds towards education and training of human resources in health institutions in order to increase the numbers of doctors, nurses and paramedics. Thus, virus coverage, shots, and cleanliness measures in the community level can help lighten hospitals during the forwarded pandemic [32].

10.17 Building Economic Resilience

Pandemic although good for the pharmaceutical industry can lead to great economic shifts, removing jobs, shuttering businesses, and severely disrupting supply lines for millions of people. As such, governments must apply strategies on aspects that limit such negative economic impacts. Making industries diverse is one of the approaches which try to avoid depending heavily on such sensitive sectors. Encouraging growth in more sectors such as e-commerce, digital finance, and renewable energy industries can be Small business should be financed through grants, low-interest loans, and other forms of access to better enable it to cope when worst comes to the worst. Studies have shown that social protection is one of the ways of ensuring that vulnerable groups are protected economically during unfavourable occurrences. Some of the interventions put in place can include payment of unemployment, direct cash transfers, and food assistance that can be essential for job losses and financial vulnerability [33]. Also, development in vocation training as well as education enables the worker to be ready for other job markets in the event that the economy changes. The other key pillar of economic readiness is the resilience of the global supply chain. The increased reliance on the "just-in-time" global value chains was ill-suited to the onset of the COVID-19 pandemic and resulted in global scarcity related to necessities such as medicines and medical gear. If countries seek to uncouple their paths from the fluctuations in the world markets, they should develop their local capacity for production and establish stockpiles of basic resources [34].

10.18 Focusing on Environmental Sustainability Promotion

It is commonly known that environmental deterioration and pandemics are related. The danger of zoonotic illnesses spreading from animals to people is increased by practices including intensive agriculture, deforestation, and wildlife commerce. A dedication to sustainable environmental measures is necessary to mitigate these hazards [13].

One of the most important ways to lessen the chance of a repeat disaster is to regulate the commerce in animals and protect ecosystems. Policies should be implemented by Governments and organizations that preserve biodiversity and prevent the exploitation of natural habitats. Additionally, sustainable agricultural practices can decrease environmental loss while ensuring food security. In addition to this, sustainable waste management is another crucial aspect of pandemic preparedness. The rise in medical waste during COVID-19, including disposable masks, gloves, and test kits, overwhelmed waste management systems in many countries across the world. Investments in facilities that can safely process hazardous waste and promote recycling thus reduce pollution and minimize health risks. Moreover, various efforts to combat climate change, such as switching to renewable energy sources and decreasing greenhouse gas emissions, are integral to long-term pandemic preparedness. Healthy ecosystems and a stable climate support public health by reducing the chances of diseases and improving air and water quality for betterment of health worldwide [29].

10.19 Focusing on Innovation and Collaboration

Pandemics are phenomena that cannot be effectively addressed by small-scale solutions. There is need for various organizations and private sectors that will participate on national as well as international level to work together for executing planning for the pandemic. Exchanging resources, knowledge, and technology can assist in improving and implementing a defence plan against health threats. Due to the advancement of Technology, it has a major role to play in boosting up pandemic readiness. Mainly, it includes use of artificial intelligence (AI), big data analytics, and genome sequencing that can improve disease surveillance, warning systems and aid the user in pervasive condition by various means [35]. Applications allow to monitor infection rates and notify the user of possible contact with a carrier have already been useful during COVID-19. Application advancement has been beneficial for the user during the time of pandemic. Further expenditure should be made towards research activities which may cause the presentation of a new generation of vaccines, drugs, and assessment tools [31].

10.20 Applying Educational and Social Justice

A pandemic always impacts vulnerable people and increases educational and social inequalities among people. During COVID-19, a lot of students suffered study loss due to non-availability of digital devices. On National and International basis there is need to properly invest in digital technologies, devices and connections to offer the opportunity and to underserved learners. Working Women, particularly those in developing countries, are usually more affected by pandemic as they have to terminate their work and have to balance home care work. There are measures that can be done to ensure equality among people by implementing various policies like equal access to all resources like financial, healthcare, education, etc. [10]. The other issue is to increase awareness for pandemic and engaging preventive measures, psychological support for those who have been affected during pandemic and engaging civil society in recovery programs. Efficient community self-organization can do collaborative action during disasters by their rapid response and action at the time of requirement according to the need of people [36].

10.21 Conclusion

COVID-19 has come as a fresh remembrance of showing how much we are interconnected with our world and how much we are exposed to disasters. The lessons that have been gained from this global pandemic is that preventing measures should be taken from a multifaceted approach that should include medical, economic and environmental security. By addressing these areas in parallel, the impact of future pandemic can be adequately mitigated. One major lesson everyone must learn from COVID-19 is the need for improvement in health systems. Several nations, particularly developing ones, poorly estimated the increase in patient number and failed to meet it as they are ill-prepared for it. Due to increasing patient number intensive care units and wards in hospitals were in high demand but due to shortage of personal protective equipment, medical personnel, oxygen, and ventilators a number of lives suffered a lot. To fill these gaps there is need of building new hospitals, improving the rural healthcare services, and applying advanced healthcare technologies. Additionally compensating and training a competent health workforce must be prioritized to support health systems when disasters occur. Another platform of pandemic preparedness is economic recovery from disruption. COVID-19 proved vulnerability of many economies, millions of people were left during lockdown due to closure of various organizations, causing the halt of various sectors. In economic development prime focus should be on the diversification of industries so that no industry is over-representative. This will prevent over – reliance on one sector. Government must also implement social protection plans to protect and support the population during crisis. Services like low unemployment benefits, food aid schemes or credit schemes for small business during these times can be of great aid. So, both health sector and economics are equally important to foster competitive local centres production and to ensure the sources supply, to guarantee the easy and steady availability of items of major necessity like medicines and food during emergencies.

Environmental sustainability plays an important role in minimizing the likelihood of pandemic and their associated risks. Many pandemics, including COVID-19, have zoonotic origins, as they arise from diseases that transfer from animals to humans. Deforestation, wildlife trade, and intensive agriculture increase the risk of these diseases. To control these risks ecosystems protection, wildlife trade regulation, and promoting biodiversity are the major steps that should be taken into consideration. Additionally, sustainable waste management systems are needed to implement to safely handle the increased medical waste like disposable masks and syringes generated during crisis, if not properly managed they can be hazardous to environment. Investments in renewable energy and green infrastructure can also contribute to a healthier planet, reducing pollution and combating climate change. Collaboration between various nations is essential for effective pandemic preparedness. The COVID-19 pandemic illustrated the power of worldwide cooperation in vaccine development and distribution. However, it also highlighted the struggle of many low-income countries' inequity in access to resources, to secure vaccines and treatments. Future efforts must focus on equitable resource distribution among various nations to establish global frameworks and to ensure that every country will receive support to overcome the critical situations. Organizations like the World Health Organization (WHO) play a crucial role in coordinating all efforts and fostering collaboration among nations. Technological innovation is another vital component of pandemic preparedness. Tools such as artificial intelligence (AI), data analytics, and genomic sequencing have already proven its important role in tracking infections, identifying hotspots, and developing vaccines. Investment in technology can further enhance and strengthen early warning systems, aid governments to response quickly and take actions during the early stages of an outbreak. Additionally, mobile health applications, telemedicine platforms, and digital health records can also improve access to care and improve overall health outcomes.

Community engagement is a foundational aspect of fostering resilience against pandemic. During COVID-19 pandemic, various organizations and government played a vital role in providing food, medical supplies, and required support to affected populations. To ensure that societies are better equipped to manage future crises strengthening these networks and empowerment of communities through education and resources is another important step. Public health initiatives that promote vaccination, hygiene, and preventive measures can also reduce the spread of diseases and promote trust in healthcare systems. Addressing social inequalities is paramount in ensuring that pandemic responses are inclusive and equitable. Vulnerable groups, including women, children, and low-income families, often suffer during health crises. It is the role of Governments to implement policies that provide targeted support to the affected populations, such as cash transfers to fulfil basic need, free healthcare services, and access to education. Bridging the digital divide is also essential, particularly as online platforms become increasingly important for education, work, and healthcare. Providing affordable internet access and digital tools to communities will ensure that everyone benefits from technological advancements. The insights gained from COVID-19 must enhance the commitment to preparedness globally. This means not only addressing immediate issues but also planning for withstanding and recovery from the disruptive events. Worldwide every sectors like Governments, international organizations, and private sectors must work together to build new systems that can respond to future shocks immediately. Investments in scientific research and development, particularly in vaccine development technology and disease surveillance, will be critical to staying ahead of emerging threats.

In conclusion, the COVID-19 pandemic has served as a wake-up call for everyone in the world. It has exposed vulnerabilities and weaknesses but also provided opportunities to build robust, more resilient systems. By adopting strategies that address healthcare, economic, and environmental challenges, communities can create a better and safer future. Preparedness is not a one-time effort; it is and continues commitment that requires collaboration, innovation, and inclusivity. The lessons gained from COVID-19 should inspire measures, ensuring that humanity must be prepared to face the challenges of tomorrow.

References

1. Labora González JJ, Fernández-Vilas E. The social evolution of COVID-19: pandemics as total social facts. Front Sociol. 2024;9:1397826.
2. Chen S, Barrett P. Social repercussions of pandemics. IMF Work Pap. 2021;2021(021):1.
3. Cordell R, Wood RM, Wright TM. Disease and dissent: epidemics as a catalyst for social unrest. Glob Stud Q. 2023;3(2):1–13.
4. Getachew E, Adebeta T, Muzazu SGY, Charlie L, Said B, Tesfahunei HA, et al. Digital health in the era of COVID-19: reshaping the next generation of healthcare. Front Public Health. 2023;11:942703.
5. Alharbi F. The use of digital healthcare platforms during the COVID-19 pandemic: the consumer perspective. Acta Inform Med. 2021;29(1):51–8.
6. Ott M, Shaw SF, Danila RN, Lynfield R. Lessons learned from the 1918-1919 influenza pandemic in Minneapolis and St. Paul, Minnesota. Public Health Rep. 2007;122(6):803–10.
7. Tomes N. "Destroyer and teacher": managing the masses during the 1918-1919 influenza pandemic. Public Health Rep. 2010;125(SUPPL. 3):48–62.
8. Kshatri JS, Satpathy P, Sharma S, Bhoi T, Mishra SP, Sahoo SS. Health research in the state of Odisha, India: a decadal bibliometric analysis (2011-2020). J Fam Med Prim Care [Internet]. 2022;6(2):169–70. Available from: http://www.jfmpc.com/article.asp?issn=2249-4863;year=2017;volume=6;issue=1;spage=169;epage=170;aulast=Faizi
9. Zaremba A, Kizys R, Aharon DY, Demir E. Infected markets: novel coronavirus, government interventions, and stock return volatility around the globe. Financ Res Lett. 2020;35:101597.
10. Ferreira FHG, Sterck O, Mahler DG, Decerf B. Inequality in the time of covid-19. LSE Public Policy Rev [Internet]. 2021;1(4):2. Available from: https://www.imf.org/external/pubs/ft/fandd/2021/06/inequality-and-covid-19-ferreira.htm
11. George TK, Sharma P, Joy M, Seelan G, Sekar A, Gunasekaran K, et al. The economic impact of a COVID-19 illness from the perspective of families seeking care in a private hospital in India. Dialogues Health. 2023;2:100139.
12. Kaye AD, Chancellor V. Since January 2020 Elsevier has created a COVID-19 resource centre with free information in English and Mandarin on the novel coronavirus COVID-19. The COVID-19 resource centre is hosted on Elsevier Connect, the company's public news and information, January. Elsevier; 2020.
13. Miyah Y, Benjelloun M, Lairini S, Lahrichi A. COVID-19 impact on public health, environment, human psychology, global Socioeconomy, and education. Sci World J. 2022;2022:5578284.
14. Ciotti M, Ciccozzi M, Terrinoni A, Jiang WC, Bin WC, Bernardini S. The COVID-19 pandemic. In: Critical reviews in clinical laboratory sciences. Taylor and Francis Ltd.; 2020. p. 365–88.
15. Jazieh AR, Kozlakidis Z. Healthcare transformation in the post-coronavirus pandemic era. Front Med. 2020;7(July):1–6.
16. Bruinen de Bruin Y, Lequarre AS, McCourt J, Clevestig P, Pigazzani F, Zare Jeddi M, et al. Initial impacts of global risk mitigation measures taken during the combatting of the COVID-19 pandemic. Saf Sci. 2020;128:104773.
17. Baig AS, Butt HA, Haroon O, Rizvi SAR. Deaths, panic, lockdowns and US equity markets: the case of COVID-19 pandemic. Financ Res Lett. 2021;38:101701.
18. Aditya B, Amri I. Rethinking informal economy resilience during crisis: experience from COVID-19 pandemic. Indian J Labour Econ. 2023;66(3):711–37.
19. Le TDQ, Nguyen DT, Ho TH, Ngo T. Government intervention and stock price returns during covid-19 pandemic: evidence from an emerging market. Cogent Bus Manag. 2024;11(1):2376042.
20. Di Fabio A. The psychology of sustainability and sustainable development for well-being in organizations. Front Psychol. 2017;8(SEP):1534.
21. Kapoor M, Nidhi Kaur K, Saeed S, Shannawaz M, Chandra A. Impact of COVID-19 on healthcare system in India: a systematic review. J Public Health Res. 2023;12(3):22799036231186349.
22. Centres OF. ISAS working papers. 2016;(226):1–18.
23. El Keshky MES, Basyouni SS, Al Sabban AM. Getting through COVID-19: the pandemic's impact on the psychology of sustainability, quality of life, and the global economy—a systematic review. Front Psychol. 2020;11:585897. Frontiers Media S.A.

24. Deshpande A. The Covid-19 pandemic and gendered division of paid work, domestic chores and leisure: evidence from India's first wave. Econ Polit. 2022;39(1):75–100.
25. Abdelrahman Z, Li M, Wang X. Comparative review of SARS-CoV-2, SARS-CoV, MERS-CoV, and influenza a respiratory viruses. Front Immunol. 2020;11:552909. Frontiers Media S.A.
26. Holshue ML, DeBolt C, Lindquist S, Lofy KH, Wiesman J, Bruce H, et al. First case of 2019 novel coronavirus in the United States. N Engl J Med. 2020;382(10):929–36.
27. Meier M, Pinto E. COVID-19 supply chain disruptions. Eur Econ Rev. 2024;162:104674.
28. Yan K, Yan H, Gupta R. The predicted trend of COVID-19 in The United States of America under the policy of "Opening Up America Again". Infect Dis Model. 2021;6:766–81.
29. Ryner JM. Silent revolution/passive revolution: Europe's COVID-19 recovery plan and green deal. Globalizations. 2023;20(4):628–43.
30. Mudera CP, Bavdekar RD, Kumar N, Veiraiah A, Nair RK. Reaching out to the millions: a 5 key messages rapid IEC campaign during the COVID-19 pandemic. Community Heal Equity Res Policy. 2022;42(3):337–41.
31. Javed S, Chattu VK. Strengthening the COVID-19 pandemic response, global leadership, and international cooperation through global health diplomacy. Health Promot Perspect. 2020;10(4):300–5.
32. Mazzucato M, Kattel R, Quaggiotto G, Begovic M. COVID-19 and the need for dynamic state capabilities: an international comparison | United Nations Development Programme. Oxford Rev Econ Policy [Internet]. 2021;36(Suppl_1):S256–69. Available from: https://www.undp.org/publications/covid-19-and-need-dynamic-state-capabilities-international-comparison#modal-publication-download
33. Alberola E, Arslan Y, Cheng G, Moessner R. Fiscal response to the COVID-19 crisis in advanced and emerging market economies. Pac Econ Rev. 2021;26(4):459–68.
34. Pujawan IN, Bah AU. Supply chains under COVID-19 disruptions: literature review and research agenda, Supply chain forum, vol. 23. Taylor and Francis Ltd.; 2022. p. 81–95.
35. Liu G. Deeper international cooperation on COVID-19 pandemic prevention and control measures in the field of migration administration, with reference to the global compact for safe, orderly and regular migration. Geneva: International Organization for Migration (IOM); 2020. p. 1–9.
36. Haelermans C, Korthals R, Jacobs M, de Leeuw S, Vermeulen S, van Vugt L, et al. Sharp increase in inequality in education in times of the COVID-19-pandemic [internet]. PLoS One. 2022; 17:1–37. Available from: https://doi.org/10.1371/journal.pone.0261114

11 The Rise of Digital Health Technology in the COVID-19 Pandemic

Anjali, Kunal Aggarwal, Mukul, and Deepanshu

11.1 Introduction

The coronavirus, which is also known as COVID-19, is a new type of virus which was discovered toward the end of the year 2019. This was a communicable viral disease spread worldwide in a very short time and caused a global emergency. On 11 Feb 2020, the International Committee on Taxonomy of Viruses called it a severe acute respiratory syndrome coronavirus 2 (SARS-COV-2). Due to the quick spread of COVID-19, the World Health Organization (WHO) announced it as a global outbreak as it caused an unusual disaster in the field of health, education, economy, cultural, etc. Global mobility was restricted at that time. To control the COVID-19 spread, the public authorities and governments of all countries imposed lockdowns, social distancing, and quarantine and stay-at-home orders. Since these orders aid the individual to stay home, use of online platforms, teleconsultation, telemedicine, remote monitoring, and health apps. Due to isolation situation, the usage of digital technologies, life protecting devices, equipment, and tools was reinforced [8]. Therefore, during COVID-19 crisis, digital health technology was the major approach.

Although digital technology was introduced decades ago, the presence of pandemic situation increases the approach of digitalization. The term digital is defined as information or any other data created by using digital platform or any device that is used to identify the test, predict disease symptoms, and prevent its future spread. The modern health tools include things like telemedicine/telehealth, electronic health records, using big data and artificial intelligence (AI) algorithms for public health surveillance, wireless health sensors, imaging, mobile health apps, and data analysis to support clinical decisions [24]. In health sector, the information technologies applied in digital manner are defined as: digital devices, digital patients, and digital clinics.

In the COVID-19 pandemic time, hospitals, societies, organizations, and other areas were supported by digital health in a more efficient manner. The digital devices are the monitoring devices which solve the clinical problems independently rather to go anywhere in clinics. The digital devices are smartphone and smartwatch,

Anjali (✉)
Department of Pediatrics, Post Graduate Institute of Medical Education and Research, Chandigarh, India

K. Aggarwal
Department of Medical Microbiology, Post Graduate Institute of Medical Education and Research, Chandigarh, India

Mukul
University Institute of Applied Management Sciences, Panjab University, Chandigarh, India

Deepanshu
Department of Nuclear Medicine, Post Graduate Institute of Medical Education and Research, Chandigarh, India

A. Raghav et al. (eds.), *COVID-19 Impact on Economy, Environment and Healthcare*,
https://doi.org/10.1007/978-981-95-1196-9_11

heart rate monitoring device, glucose, oxygen, body temperature monitoring, wireless as well as wearable devices and implant sensors. These devices can easily be used by an individual at home. A digital patient is defined as a population which follows the mobile health (m Health) technologies including patient self-measurements by mobile app and digital devices and also telemedicine to manage their health. The main focuses of the digital clinics are smartphone-based tracking, virtual hospitals, remote patient monitoring and for generating mobile health data, analyzing it to make it clinically useful and incorporate into regular healthcare practices.

In the pandemic time World Health Organization define digital health like a spider or as a broad umbrella in health aspects which consist of electronic health also term as mobile health (e-health) along with new fields such as the application of advanced computing sciences in "big data," genomics, and artificial intelligence and robotic health [18]. Digital technology divides among developed and developing nations, urban and rural populations; economic and cultural differences make healthcare services more challenging and difficult [8]. The use of digital health technology needs more current and accurate high-quality data information with strong trust between the consumers and healthcare service providers, because of uncertainty around COVID-19. Thus, digital healthcare platforms are also referred to as collections of applications and technologies used to support the delivery of healthcare services [2].

The pandemic situation in the worldwide has become a strong factor for enhancing the adoption and implementation of digital tools to reduce the transmission of COVID-19. It also provides an easy platform for health care providers by which people can receive medical advice and care via online mode and can consult to physicians, monitor their health and get treatments without leaving their homes [2].

11.2 Digital Technologies of Healthcare and Its Objectives

The term digitalization is characterized as the utilization of digital technology to update the business model and is also known as the "ability to turn existing products or services into digital variants." On the other hand, digitalization refers to the "adoption or increase in use of digital or computer technology by an organization, industry, etc." [21]. This is a technical process of converting analogs, paperwork into the digital form so that the digital devices such as computer, laptop, smartphones, etc. can help in accessing, sorting, and transmitting the information and is also termed as a sociotechnical process of utilizing digitized products or system to develop new organizational procedures, business model or commercial offerings. The major goal of digitalization is to build a relationship between the service provider and the customers.

During COVID-19 pandemic situation, in a short period of time, an accelerated use of digitalization took place in all countries in each and every field of life. A study by Amankwah-Amoah et al. [3] explains how COVID-19 was the "greatest accelerator" and acted as "catalyst" for digitalization. Around 37% of the jobs in the United States were converted home based. Video-conferencing and remote collaborations increase the remote work process, reducing human error.

During the pandemic situations (SARS-CoV-2 virus, Ebola virus, Zika virus), the direct interaction of the patient and a doctor was replaced with digital technology. This also minimizes the cost of travel, time of the individual, and also reduces unemployment. This was a valuable tool for the patient's convenience. Digitalization probably has a meaningful potential to raise health awareness and improve preventive measures. According to the World Health Organization (WHO) and

United Nations Development Programme (UNDP), digital technologies can improve patient care and health outcomes by enabling better diagnoses, data-driven treatment choices, digital therapies, clinical trials, self-care, and patient-focused services. These technologies can also help health centers increase revenue, provide more accurate data for policy decisions, lower costs for patients, and support health workers' learning through e-learning. This approach aims to upgrade patient safety and quality of care.

Key objectives of digital health: World Health Organization explains the objectives of the digital health, some are as given below:

I. To convert both existing and newly generated dataset into action that informs and guides decision-making process.
II. The application of digital technologies to enhance connectivity and information transfer, sharing, including remote communication.
III. The systematic assessment of the national and regional needs for emerging new technologies, along with promoting technological co-development.

Digital health-related solutions address the societies how modern technology of health care system can support human health by tracking, screening, identification, diagnosis, monitoring, and contact tracing; no data provided on the use of this approach in practice for fighting COVID-19. Thus digital technology has great potential for controlling the COVID-19 outbreaks and should be considered at the top level of health systems; they also discussed the challenges that policymakers might face in controlling the crisis using digital solutions [19].

Digital technologies in healthcare explain how health services are being provided and how the overall healthcare service operates. Data management has the main focus for digital healthcare. The latest technologies such as artificial intelligence (AI) and machine learning tools assist the healthcare sector in analyzing and assessing vast amounts of patient data [22].

11.3 Monitoring of COVID Using Digital Technology

Digital technologies in health sectors are rising day by day. In COVID-19 eras, digital health technologies which are used in health sectors are:

11.3.1 Electronic Health Record (EHR) Devices and e-Prescriptions

The COVID-19 pandemic caused sudden and major changes in many fields of the healthcare delivery system, including clinician workflows and use of electronic health record systems [14]. The Electronic Health Record (EHR) is an electronic record of patient health information like demographics, progress notes, problems, medications, vital signs, past medical history, immunization, laboratory, and radiological data. It improves the documents record keeping of patients and data integrity of the healthcare services. This is the easy way to share the patient data without late. Electronic health records (EHR) can help to maintain, sort, store, and process the clinical data. Moreover, it also reduces the cost of healthcare and number of medical errors in clinical organization. In developed countries the e-health has already developed and officially approved platforms to receive and analyze medical information. Electronic documentation system can accumulate the information about the patient throughout the life. E-health systems enable doctors to write and send electronic prescriptions directly to the pharmacy, which will also significantly reduce the patient wait time in clinics and paperwork. HER also improves healthcare productivity.

11.3.2 Telehealth/Telemedicine

Telehealth which is also called as telemedicine is a healthcare service provided through telecommunications (Radio and telephone). Telemedicine allows communications between patient and a healthcare professional through social media communication platforms like video apps such as Viber, Google Meet, Microsoft Teams, Facebook, Zoom using desktop, laptops, and smartphones. "Community Medical Education" (CMED) is a health organization that created a digital health service for COVID-19 outbreak that includes education, awareness, screening, triage for possible cases, sample collection, lab tests, and patient management [8]. It provides the elective care. Telemedicine can help people in rural, remote, and underserved areas access healthcare more easily.

Telehealth services are divided into three categories, i.e., synchronous, asynchronous, and remote patient monitoring [13]. In synchronous services, there is a direct connection between the healthcare provider and patient; in these services, the way of data transmission is done by using phone, video, and texting. Asynchronous services allow patient and healthcare providers to store information. On the other side, remote patient monitoring (RPM) is the mixing of both communications: synchronous and asynchronous. Remote patient monitoring evaluates the progress of patient over time.

The American Telemedicine Association (The ATA) describes RPM as "home telehealth" in which data is remotely collected and sent to health agencies for interpretation [13]. Thus telemedicine is the time-saving method for diagnosis and treatment startup. With the help of telemedicine intra-hospital infection and the cost of antiseptic materials can be reduced. It also reduces the direct physical contact which will ultimately reduce the risk of respiratory infections. Training of health cares, physicians, or professionals can be done by telemedicine and monitoring of real-world data can be done by various tools of telemedicine.

Telehealth (e-health) is also classified on the basis of features and its functionalities are:

(a) Services offered without patient self-measurement data.
(b) Exchange of information and communication related to well-being of the patient.
(c) Monitoring and managing the clinical data of the patient to aid in diagnosis of disease and helps in clinical decision.

The most common methods of using telemedicine are real-time technology, store-and-forward technology, remote monitoring, and M-health approaches. In Ethiopia, Shiferaw et al. [25] conducted an online survey to find the telemedicine modalities which showed 36% of respondents prefer m-heath, 28% showed interest in real-time communication (live streaming), 12% use remote patient monitoring services (RPM), and 24% record and send (store and forward) [8].

11.4 Telehealth Applications During the COVID-19 Pandemic

During the pandemic time virtual treatment was followed by all societies. Further in COVID, due to telehealth, it is easier for patients to access their health information and communicate with clinicians online, leading to increases use of electronic health records. EHR data can play a crucial role in disease monitoring and exposure tracking; assessing patient risk level and real-time monitoring, early detection and management of disease severity; identify risk factors and disease complications, identify persistent outcomes; and also offering a deeper understanding of disease mechanisms through advanced molecular techniques (genomics, metabolomics and proteomics) and also helps to create advanced tools like artificial intelligence, remote monitoring for early detection of pandemics [24]. This is a reasonable, productive, and attractive option for the societies. This also improves individual safety. Before the COVID pandemic, telecare already existed in developed countries like the United States, Canada, but during pandemic the use of telecare became wide in all world. According to Bouabida et al. [7] in October 2020 there was 13,000%

growth of telehealth worldwide. They also mentioned the use of online consultations has increased. Before COVID the use of online consultation was 19.3% and after pandemic 41.2%.

Various mobile app has been developed at the time of COVID-19 outbreak to track the individual information to overcome the disease spread. An online application connected to the EHR systems send an alert when individual tests positive for COVID. With the help of telehealth infection spread rate become lower by online tracking the disease symptoms, enhancing the social distancing. It also provides quality care of the individual.

Telehealth can provide an important resource to provide suitable health interventions and outcomes of human health. During the global crisis like COVID-19, Zika virus the rise of digital health (telehealth) in time was the promising approach in the improvement of healthcare system. There are some examples of use of telemedicine services in India at the time of COVID-19, some are—"eSanjeevani OPD" which is a nationwide remote consultation service provided by Indian government, it provides healthcare support for patients staying at home. Other is "Practo" which is an online application that provides an appointment booking experience to the patients through which they can easily book their appointment with relevant healthcare professional (doctors, nurses, laboratory technicians, and others). India's health care brand (Foreign OPD) provides face-to-face consultation with doctors and medical health experts worldwide [8] (Table 11.1).

Telepharmacy services (Remote prescription): Telepharmacy services also played an important role in controlling the pandemic situation. The digital health approach could become the main method for prescribing or refilling medications for chronic illness patients. A study showed an 88% satisfaction and acceptance rate for remote prescriptions among these patients. Another study with 52 community pharmacies found that remote prescription services are effective in reducing strain on the healthcare system during the pandemic [1]. However, it showed during pandemic remote prescriptions are the effective tools to minimize the pressure on healthcare system. Thus, digital health acts as an empowering tool for the healthcare system during the pandemic situation.

Table 11.1 Tele application used in COVID-19 to control the situation

Tele application	Objectives	Service provided to	Country
eSanjeevani OPD	Booking appointment	Patients, doctors, and healthcare providers	India
Practo	For booking online consultation	Healthcare personnel and patient as well	India
Sneha	Book online appointments to maintain health records and check daily sign & symptoms	Healthcare providers patients, doctors	Saudi Arabia
Arogya setu	Daily sign and symptoms & to maintain real-time data	Case and societies	India
Head to toe	Provide online education and coronavirus-related information and practice	Healthcare providers	Switzerland
National health Sewa	Control social distance, analyzing test results, and data sharing	Policymakers, healthcare providers and patients	UK

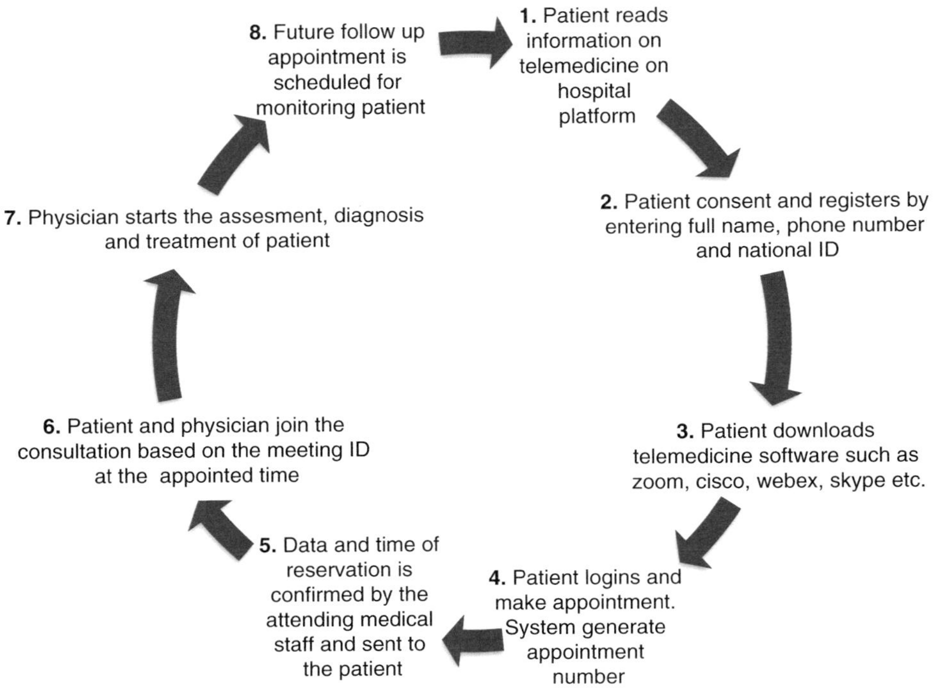

Flow diagram describes the workflow of the telehealth

11.5 Artificial Intelligence in Healthcare: A Major Approach

Artificial intelligence is an evidence-based medical digital technology which helps in the diagnosis of various diseases more accurately, quickly, and reliably which is ultimately useful in predicting the disease pattern and epidemiology in a real-time manner. AI is considered as a foundation of digital technology. It is an innovative technology comprised of (i) component: like hardware (machine and devices) and software devices (programs and algorithm), (ii) technology: natural language processing (helping computers to respond to human languages, machine learning (computer learn from data)), (iii) medical imaging (CT scans, MRI, radiomics, PET-CT, X-rays, etc.), robotics (machines that can do tasks), smartphones, smart wearables, automobiles, and security systems, (iv) vertical (healthcare, advertising and media, telecommunication, etc.) [8]. It is a smart and potential tool which removes the manual health system into automatic by minimizing the errors caused by human negligence. Artificial intelligence is widely used for preparing treatment plans, analyzing medical records, drug development, for detecting hotspots and contact tracing [20]. It also helps in various healthcare systems, including diagnosis, therapeutics, clinical decision-making, and public health. It also predicts the mortality risk by analyzing the previous data of the patient. AI is a novel technology which works more efficiently in medical healthcare system. It is better than human for performing healthcare tasks and improving the human in a cost-effective manner.

In clinical practice, AI is also called as clinical decision support systems (CDSS) which provides support to clinician in treatment decision of various kinds of diseases like cancer, tuberculosis, cardiac, diabetes, and neurological diseases, etc. [4]. The invention of artificial intelligence tools reduces workload of individual but also increases the quality care of patient care, and reduces unnecessary hospital admission. The effect of artificial intelligence in healthcare system can be explained in four ways which are: helping, dividing tasks, replacing jobs, and improving work. Economically, it saves time and resources.

This is a preliminary tool to scan radiological images, radio-nucleotide image, screen them, and completing the report for approval by physician. By using AI tools medical errors can be reduced which results in better clinical decision-making by providing up-to-date clinical guidelines, policies, and development, which may increase the medical knowledge. This is a promising approach for clinical genomics analysis, including genome annotation, variant classification, and phenotype-genotype communications [5].

11.6 Types of Artificial Intelligence

Artificial intelligence is not one technology but also it is a collection of various tools which widely provide support to healthcare. The tools of the artificial technology which involved in healthcare are: Machine learning, physical robots, and robotic process automation.

11.6.1 Machine Learning and Deep Learning

This is the most common tool of AI. It is a statistical and broad technique that provides technical support for early detection and diagnosis of disease by using computational models (CT scans, X-rays, and ultrasound images) on large clinical data. Machine learning, also called as "neural network", is used in categorization application to determine whether a patient acquires a particular disease. The complex form of machine learning is deep learning which describes the radiomics data, detection of relevant clinical features in imaging data. This is an automated tool for image analysis [11].

11.6.2 Physical Robots

A machine which can perform multiple tasks to enhance the patient care and has high autonomy than humans. In the digital technology, robotic tools are the recent and advanced technology to increase productivity. This is a type of digital innovation used in healthcare. The term robot originates from the Czech word "robota" which means slave, servant, or forced labor [26]. Robots are defined as mechanical artificially intelligent agents that can perform multiple tasks and ability to move. Robots can work in automatic or manual way. They can sense things and carry out the computation to make decisions and perform mechatronic actions. Robots can be classified as: service robots and industrial robots. Service robots are also called physical robots which can work as human beings and are used for personal, professional settings. Alternately, the industrial robots are used in industrial work in production and manufacturing.

11.6.3 Robotics in Healthcare

Service robots are used in healthcare; they have direct interaction with human users in hospitals, rehabilitation clinics, and the family home. They provide support to the clinical staff, functioning of impaired individuals, medical intervention, and social assistance, and disinfection, interventional and imaging assistance and in rehabilitation. They have direct interaction with human users and provide both physical and cognitive support. A robot in the healthcare laboratories does outstanding productive tasks and reduces manual work and also improves efficiency. Mobile robots in healthcare also support sanitation services.

During COVID-19 pandemic, mobile robots are used to minimize the spread of infection in hospitals, minimize the human-human contact, and act as safeguard. Autonomous robot during the COVID-19 pandemic can manage and monitor people's movements, ensuring that they follow lockdown rules and avoid unnecessary gathering movement. During COVID-19 tele-robots are used to do the surgical procedures, diagnosis and treatment. In 2000, the surgical robot was the major invention which was approved by the United States, which provides s superpower to the surgeons by improving their ability to see the disease location, minimize the invasive incisions, stitch the wound, laparoscopic surgery including prostatectomy, hysterolaproscopy, cholecystectomy, pancreatectomy and head and neck surgery, etc. [12]. Robotic groups have a well-defined area of use in healthcare sectors.

11.7 Application of AI in COVID Era

During COVID-19 pandemic to date 15 January 2021, in 191 countries and regions, a total of 93,129,104 confirmed cases and 1,994,440 deaths were recorded [23]. During the pandemic time many researchers were seeking various technologies to suppress the virus spread. Between them the artificial intelligence was a potential powerful data analytical tool in fighting against the disease. This technology is a highly efficient and reliable approach which reduces time for epidemic prevention and reduces the mortality.

The main applications of artificial intelligence in the COVID-19 pandemic are:

11.7.1 Tracking and Prediction of the Disease

During the COVID-19 outbreak artificial intelligence techniques are used to track and predict the disease spread. AI-based model from HealthMap at Boston Children's Hospital in the United States raised an early alert on December 30, 2019, approximately 30 min ahead of a similar notification issued by a scientist from the Program for Monitoring Emerging Diseases (PMED) [17]. Big data can monitor and track the spread of disease in real-time as it happens. Using LSTM, network prediction of coronavirus spread can be done. LSTM and curve-fitting were used by Tomar and Gupta to calculate and predict the number of COVID recovered cases, number of daily positive new cases, number of death rate of cases, and their preventive actions. Alternately in 2020 Chen et al. [9] used LSTM to predict the spread of coronavirus [23]. Bogu and Snyder proposed in the study that data from wearable sensors could help in early prediction of COVID-19. With real-time data from wearables, such as heart rate or body temperature, researches can track signs of the virus. The combined information with the lab tests can improve our understanding of how to detect and monitor COVID-19 outbreaks more effectively.

11.7.2 Disease Diagnosis and Prognosis

Artificial intelligence plays an important role in the fast detection of disease using AI algorithm and medical imaginary. The methodologies used to detect COVID-19 are (i) The Deep Neural Networks (DNNs), (ii) Hand-Crafted Feature Learning (HCFL), and (iii) the hybrid approach. With the help of artificial intelligence medical imaging technologies like X-rays, computed tomography (CT scan), magnetic resonance imaging (MRI), the diagnosis of COVID-19 becomes fast and accurate. This helps in saving the life of the person in COVID time. Machine learning and big data can help to predict the disease progression. Artificial intelligence helps in proper monitoring and treatment of the affected individuals. With the help of AI day to day updates of the patients and solution to overcome the disease spread are provided in the COVID-19 pandemic.

11.7.3 Development of Treatment

During COVID outbreak, AI helps to design diagnostic tools and in treatment development. A study of Google's DeepMind, which is known for its

AlphaGo algorithm, used AI to predict the structure of the virus's proteins, which could help in developing new drugs. However, as DeepMind states these predictions have not been experimentally verified, so the accuracy of the structures cannot be guaranteed [16]. Beck et al. [6] discovered that the drug atazanavir could be a potential drug used to treat COVID-19 infection. In this study machine learning model was used by Beck et al. [6]. Similarly, Stebbing et al. identified Baricitinib, a drug used for rheumatoid arthritis and myelofibrosis, as a possible treatment for COVID-19.

11.7.4 Reduce Mortality Risk

Machine learning model, an advanced AI tool, helps to develop a prognostic prediction algorithm to predict mortality risk of the infected person. In Tongji Hospital in Wuhan, China, Jiang et al. developed an AI model that predicts with 80% accuracy which individuals with COVID-19 are likely to develop acute respiratory distress syndrome (ARDS) [17]. To predict mortality risk of infected patient "prognostic prediction algorithm" was developed by Yan et al. [27] by using machine learning model.

11.7.5 Survival Prediction

With the help of AI, survival prediction can be done. A group of scientists [27] conducted a study in China to predict survival in COVID-19-infected patients. In this study, they used a clinical database of blood sample from infected subjects to predict the disease severity. By selection of three biomarkers (lymphocytes, C-reactive protein, and lactate dehydrogenase) in the machine learning model (ML), Yan et al. calculate the survival rate of the infected individual. They found the accuracy to be 90%. This shows that with the help of blood biomarkers, one can predict the survival of the COVID-infected patient [23]. Various ML models have been used successfully to detect COVID-19, predict patient outcomes, and assist in early detection, even without relying on advanced imaging like CT scans. These models are showing great promise in improving accuracy and aiding healthcare professionals in managing the pandemic.

11.7.6 Reducing the Workload of Healthcare

During COVID time, there was a massive increase of patients hence workload of the healthcare increased. Digital platforms of artificial intelligence models (teleheath, remote patient monitoring, and telemedicine) are used to reduce the workload of the healthcare workers.

11.7.7 Social Control or Contact Tracing

Artificial intelligence tool of thermal scanning is used to scan public spaces which help to identify the infected patient and by enforcing social distancing and lockdown measures. Biosensors are used in pandemic time for recording human body temperature, heart rate, tracing the electrocardiogram (ECG), and also help in detecting COVID-19 symptoms in real-time. This helps to encourage the individual to stay a home and minimize hospital visits which ultimately enhance social distancing. For lockdown monitoring, thermal imaging, medical supply, public announcements, disinfection, and surveillance IoT-equipped drones are used. This helps to maintain social distancing.

11.7.8 Prevention of Disease

With the help of real-time data analysis, AI is helpful in the prevention of disease by early detection of virus using AI tools. By predicting viral biomarkers, imaging techniques, thermal monitoring devices, etc. AI helps in the prevention of the viral disease.

11.7.9 Improve Quality Care of the Subjects

Artificial intelligence also helps to improve the quality care of the subjects. Due to quarantined situation, persons get their treatment remotely which reduces the social distancing and time of hospital visit for treatment. Work from home (WFH) in COVID-19 leads to improve the quality care of the societies. AI-based social media and messaging platforms (Twitter, WhatsApp, Instagram, Facebook, etc.) in COVID-19 observed how people are reacting to the pandemic, how they are seeking health information, and what behaviors they adopt such as mask-wearing or social distancing. These trends aim to improve the health responses and communication between humans.

11.7.10 Big Data Analysis

This is an analytical tool to extract the detailed information of complex data set of the healthcare. The analysis of Big data using AI tool can be used in the COVID-19 outbreak in data management and for tracing the history of infected people. Computer languages like Python with its NumPy are the most renowned approach in the domain of Big Data [8]. A 3D CNN network is used by Oyang et al. to diagnose COVID-19-infected patients; alternately, Wang et al. used CNN-based deep learning to screen for COVID-19 in patients finding the accuracy of 89.5%, sensitivity 87%, and specificity of 88% using CT images.

11.7.11 Application Development

Artificial intelligence technology is used in the development of software and programs against COVID-19 in epidemic prevention and control. AI develops a "Neural network" and DNN (deep neural networks) technology in which imaging pattern (like X-rays or CT scans) recognition is done and used in symptoms recognition. AI helps interpret the imaging pattern from different scans, can explain the damage caused by the virus, and also highlight patterns of infection.

11.7.12 Administrative Application

AI has great administrative applications in health sectors. Robotic process automation (RPA) can be used for clinical documentation, medical records managements, claims processing, and revenue cycle management. The healthcare professionals (physicians and nurses) could save their time using voice text writing. Additionally, chatbots have also been used by healthcare for mental health, telehealth, and patient interfaces. AI tools have a great value for clinical pharmacy services in real world. Corny et al. [10] state in a study that a hybrid ML-based decision supports the health care system more accurately in detecting the prescribing errors. Electronic health record systems are also widely used. Li et al. [15] establish the bidirectional encoder representations from a transformer for HER (BEHRT). The BEHRT is a deep neural sequence transduction model for EHRs. It can integrate various heterogeneous concepts such as assessment and diagnosis. It enhances the accuracy of the prediction.

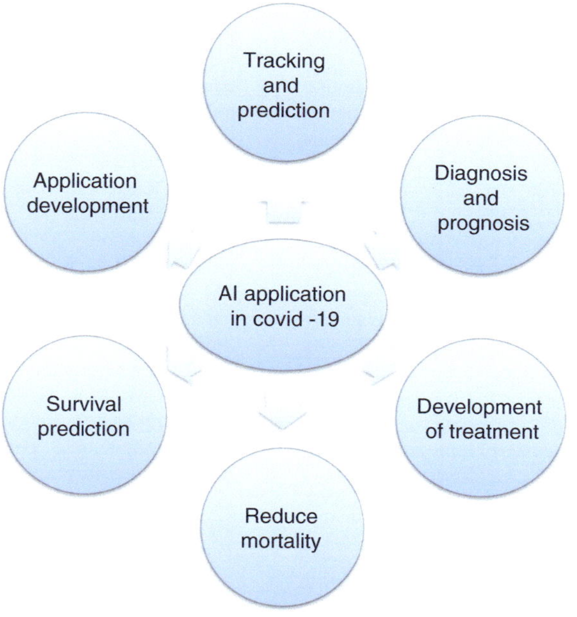

Flow diagram for the role of artificial intelligence in COVID-19

11.8 Conclusion

The impact of COVID-19 on healthcare services and society during the 2020 period was vast and numerous that initiated the rise of digital revolution that in turn had positive impact on the healthcare systems worldwide. The rise of digital health approaches including telehealth, telemedicine, and artificial intelligence during pandemic time was proven to be essential tools to mitigate the transmission of the virus. Digital innovations can improve healthcare accessibility, quality, and flexibility, benefitting both developed and developing countries. However, improvement in digital technologies makes health services smarter and can reduce long-term healthcare cost by improving precision and reducing errors ensuring better quality and availability of care. As the need of digital technologies continues to grow, societies will need to create rules and systems for managing their development that benefit everyone and avoid potential problems. Hence, the growth of digitalization during COVID-19 has a clear potential to improve the experience for both patients and doctors and its application for future prospective will contribute to the regulation of the healthcare system efficiently.

References

1. Alghamdi NS, Alghamdi SM. The role of digital technology in curbing COVID-19. Int J Environ Res Public Health. 2022;19(14). MDPI https://doi.org/10.3390/ijerph19148287.
2. Alharbi F. The use of digital healthcare platforms during the COVID-19 pandemic: the consumer perspective. Acta Inform Med. 2021;29(1):51–8. https://doi.org/10.5455/AIM.2021.29.51-58.
3. Amankwah-Amoah J, Khan Z, Wood G, Knight G. COVID-19 and digitalization: the great acceleration. J Bus Res. 2021;136:602–11. https://doi.org/10.1016/j.jbusres.2021.08.011.
4. Amann J, Blasimme A, Vayena E, Frey D, Madai VI. Explainability for artificial intelligence in healthcare: a multidisciplinary perspective. BMC Med Inform Decis Mak. 2020;20(1) https://doi.org/10.1186/s12911-020-01332-6.
5. Aung YYM, Wong DCS, Ting DSW. The promise of artificial intelligence: a review of the opportunities

and challenges of artificial intelligence in healthcare. Br Med Bull. 2021;139(1):4–15. Oxford University Press. https://doi.org/10.1093/bmb/ldab016.
6. Beck BR, Shin B, Choi Y, Park S, Kang K. Predicting commercially available antiviral drugs that may act on the novel coronavirus (2019-nCoV), Wuhan, China through a drug-target interaction deep learning model. Comput Struct Biotechnol J. 2020;18:784–90. https://doi.org/10.1101/2020.01.31.929547.
7. Bouabida K, Lebouché B, Pomey MP. Telehealth and COVID-19 pandemic: an overview of the telehealth use, advantages, challenges, and opportunities during COVID-19 pandemic. Healthcare (Switzerland). 2022;10(11) https://doi.org/10.3390/healthcare10112293.
8. Chandra M, Kumar K, Thakur P, Chattopadhyaya S, Alam F, Kumar S. Digital technologies, healthcare and Covid-19: insights from developing and emerging nations. Heal Technol. 2022;12(2):547–68. Springer Science and Business Media Deutschland GmbH. https://doi.org/10.1007/s12553-022-00650-1.
9. Chen J, Wu L, Zhang J, Zhang L, Gong D, Zhao Y, Hu S, Wang Y, Hu X, Zheng B, Zhang K, Wu H, Dong Z, Xu Y, Zhu Y, Chen X, Yu L, Yu H. Deep learning-based model for detecting 2019 novel coronavirus pneumonia on high-resolution computed tomography: a prospective study. Sci Rep. 2020;10(1):19196. https://doi.org/10.1101/2020.02.25.20021568.
10. Corny J, Rajkumar A, Martin O, Dode X, Lajonchère JP, Billuart O, Bézie Y, Buronfosse A. A machine learning-based clinical decision support system to identify prescriptions with a high risk of medication error. J Am Med Inform Assoc. 2020;27(11):1688–94. https://doi.org/10.1093/jamia/ocaa154.
11. Davenport TH, Kalakota R. The potential for artificial intelligence in healthcare. Future Healthc J. 2019;6(2):94–8. https://doi.org/10.7861/futurehosp.6-2-94.
12. *Davenport Thomas 2019 the potential of healthcare*; n.d.
13. Hoffman DA. Increasing access to care: telehealth during COVID-19. J Law Biosci. 2020;7(1) https://doi.org/10.1093/jlb/lsaa043.
14. Holmgren AJ, Downing NL, Tang M, Sharp C, Longhurst C, Huckman RS. Assessing the impact of the COVID-19 pandemic on clinician ambulatory electronic health record use. J Am Med Inform Assoc. 2022;29(3):453–60. https://doi.org/10.1093/jamia/ocab268.
15. Li Y, Rao S, Solares JRA, Hassaine A, Ramakrishnan R, Canoy D, Zhu Y, Rahimi K, Salimi-Khorshidi G. BEHRT: transformer for electronic health records. Sci Rep. 2020;10(1) https://doi.org/10.1038/s41598-020-62922-y.
16. Naudé W. Artificial intelligence against COVID-19: an early review; 2020a. www.iza.org
17. Naudé W. Artificial intelligence vs COVID-19: limitations, constraints and pitfalls. AI & Soc. 2020b;35(3):761–5. https://doi.org/10.1007/s00146-020-00978-0.
18. Ndayishimiye C, Lopes H, Middleton J. A systematic scoping review of digital health technologies during COVID-19: a new normal in primary health care delivery. Heal Technol. 2023;13(2):273–84. Springer Science and Business Media Deutschland GmbH. https://doi.org/10.1007/s12553-023-00725-7.
19. Niakan Kalhori SR, Bahaadinbeigy K, Deldar K, Gholamzadeh M, Hajesmaeel-Gohari S, Ayyoubzadeh SM. Digital health solutions to control the COVID-19 pandemic in countries with high disease prevalence: literature review. J Med Internet Res. 2021;23(3). JMIR Publications Inc https://doi.org/10.2196/19473.
20. Nirmala AP, More S. Role of artificial intelligence in fighting against covid-19. In: Proceedings of 2020 IEEE international conference on advances and developments in electrical and electronics engineering, ICADEE 2020, December 10; 2020. https://doi.org/10.1109/ICADEE51157.2020.9368956.
21. Parviainen P, Tihinen M, Kääriäinen J, Teppola S. Tackling the digitalization challenge: how to benefit from digitalization in practice. Int J Inf Syst Proj Manag. 2017;5(1):63–77. https://doi.org/10.12821/ijispm050104.
22. Paul M, Maglaras L, Ferrag MA, Almomani I. Digitization of healthcare sector: a study on privacy and security concerns. ICT Express. 2023;9(4):571–88. Korean Institute of Communications and Information Sciences. https://doi.org/10.1016/j.icte.2023.02.007.
23. Peng Y, Liu E, Peng S, Chen Q, Li D, Lian D. Using artificial intelligence technology to fight COVID-19: a review. Artif Intell Rev. 2022;55(6):4941–77. https://doi.org/10.1007/s10462-021-10106-z.
24. Scott BK, Miller GT, Fonda SJ, Yeaw RE, Gaudaen JC, Pavliscsak HH, Quinn MT, Pamplin JC. Advanced digital health technologies for COVID-19 and future emergencies. Telemed E-Health. 2020;26(10):1226–33. https://doi.org/10.1089/tmj.2020.0140.
25. Shiferaw KB, Mengiste SA, Gullslett MK, Zeleke AA, Tilahun B, Tebeje T, Wondimu R, Desalegn S, Mehari EA. Healthcare providers' acceptance of telemedicine and preference of modalities during COVID-19 pandemics in a low-resource setting: an extended UTAUT model. PLoS One. 2021;16(4 April 2021) https://doi.org/10.1371/journal.pone.0250220.
26. Silvera-Tawil D. Robotics in healthcare: a survey. SN Comput Sci. 2024;5(1) https://doi.org/10.1007/s42979-023-02551-0.
27. Yan L, Zhang H-T, Goncalves J, Xiao Y, Wang M, Guo Y, Sun C, Tang X, Jin L, Zhang M, Huang X, Xiao Y, Cao H, Chen Y, Ren T, Wang F, Xiao Y, Huang S, Tan X, et al. A machine learning-based model for survival prediction in patients with severe COVID-19 infection; 2020. https://doi.org/10.1101/2020.02.27.20028027.

MIX
Papier aus verantwortungsvollen Quellen
Paper from responsible sources
FSC® C105338

If you have any concerns about our products,
you can contact us on
ProductSafety@springernature.com

In case Publisher is established outside the EU,
the EU authorized representative is:
Springer Nature Customer Service Center GmbH
Europaplatz 3, 69115 Heidelberg, Germany

Printed by Libri Plureos GmbH
in Hamburg, Germany